Constantine Kreatsoulas

Theoretical Models
of Skeletal Muscle

'... ἃ μὴ οἶδα οὐδὲ οἴομαι εἰδέναι.'

ΠΛΑΤΩΝΟΣ ΑΠΟΛΟΓΙΑ ΣΩΚΡΑΤΟΥΣ

Theoretical Models of Skeletal Muscle

Biological and Mathematical Considerations

Marcelo Epstein
Department of Mechanical Engineering, The University of Calgary, Canada

and

Walter Herzog
Faculty of Kinesiology, The University of Calgary, Canada

JOHN WILEY & SONS
Chichester • New York • Weinheim • Brisbane • Singapore • Toronto

Copyright © 1998 by John Wiley & Sons Ltd,
Baffins Lane, Chichester,
West Sussex PO19 1UD, England

National 01243 779777
International (+44) 1243 779777
e-mail (for orders and customer service enquiries): cs-books@wiley.co.uk
Visit our Home Page on http://www.wiley.co.uk or http://www.wiley.com

All rights reserved. No part of this publication may be reproduced, stored in a retrieval system, or transmitted, in any form or by any means, electronic, mechanical, photocopying, recording, scanning or otherwise, expect under the terms of the Copyright, Designs and Patents Act 1988 or under the terms of a licence issued by the Copyright Licensing Agency, 90 Tottenham Court Road, London, W1P 9HE, UK, without the prior permission in writing of the publisher.

Other Wiley Editorial Offices

John Wiley & Sons, Inc., 605 Third Avenue,
New York, NY 10158-0012, USA

WILEY-VCH Verlag GmbH, Pappelallee 3,
D-69469 Weinheim, Germany

Jacaranda Wiley Ltd, 33 Park Road, Milton,
Queensland 4064, Australia

John Wiley & Sons (Asia) Pte Ltd, 2 Clementi Loop #02-01,
Jin Xing Distripark, Singapore 129809

John Wiley & Sons (Canada) Ltd, 22 Worcester Road,
Rexdale, Ontario M9W 1L1, Canada

Library of Congress Cataloging-in-Publication Data
Epstein, M. (Marcelo)
   Theoretical models of skeletal muscle / Marcelo Epstein and Walter Herzog.
      p.   cm.
   Includes bibliographical references and index.
   ISBN 0-471-96955-9 (alk. paper)
   1. Striated muscle–Mechanical properties.   2. Striated muscle-
-Mathematical models.   I. Herzog, W. (Walter), 1955-   .
II. Title.
   [DNLM: 1. Muscle, Skeletal–physiology.   2. Models, Biological.
WE 500 E64t 1998
QP321.E64   1998
611'. 0186–dc21
DNLM/DLC
for Library of Congress   97–42733
                                                                   CIP

British Library Cataloguing in Publication Data

A catalogue record for this book is available from the British Library

ISBN 0 471 96955 9

Typset in 10/12pt Times by Mathematical Composition Setters Ltd, Salisbury, Wiltshire.
Printed and bound in Great Britain by Biddles Ltd, Guildford and King's Lynn.
This book is printed on acid-free paper responsibly manufactured from sustainable forestry in which at least two trees are planted for each one used for paper production.

Contents

Preface	ix
Introduction	xi

Part One
Foundations

1	**Basic introduction to skeletal muscle**	**3**
	1.1 General considerations	3
	1.2 Muscle structure	3
	1.3 Muscle force control	9
	1.4 Energy considerations	16
	1.5 Types of contractions	18
	1.6 How does a muscle produce force?	19
	Problems	20
	References	21
2	**Modelling skeletal muscle using simple geometric shapes: biological considerations**	**23**
	2.1 Introduction	23
	2.2 Shapes	24
	2.3 Contractile element properties	25
	2.4 Angle of pinnation	48
	2.5 Passive element properties	52
	Problems	57
	References	64
3	**Hill and Huxley type models: biological considerations**	**70**
	3.1 Introduction	70
	3.2 Hill type models	70
	3.3 Huxley (cross-bridge) type models	75
	Problems	84
	References	84

4 Rheological and structural models: mathematical considerations — 87
4.1 Rheological models — 87
4.2 Structural models — 101
Problems — 109
References — 110

Part Two
Applications

5 Fundamentals of mechanics — 115
5.1 Introduction — 115
5.2 Coordinates, displacement, and elongation — 118
5.3 Rates and virtual displacements — 123
5.4 External and internal forces — 124
5.5 The constitutive equation — 127
5.6 The principle of virtual work — 128
5.7 Example: a one-degreee-of-freedom system — 131
5.8 A more general example — 134
5.9 Geometric constraints — 136
5.10 Example: preservation of area — 138
Problems — 144
References — 144

6 Towards a complete muscle model — 145
6.1 Introduction — 145
6.2 A program for static analysis of skeletal muscle — 146
6.3 Example: static deformation of a cat medial gastrocnemius muscle — 148
6.4 Time-dependent modelling — 150
6.5 A program for time-dependent analysis of skeletal muscle — 151
6.6 Example: time-dependent deformation of a cat medial gastrocnemius muscle — 152
References — 152

7 Movement control — 153
7.1 Introduction — 153
7.2 The neurophysiology of movement control — 153
7.3 The anatomy of movement control — 157
7.4 Theoretical and experimental considerations on movement control — 160
7.5 Future research in the area of mechanics in movement control — 171
References — 172

Appendix A: Topics in time-independent modelling — 174
A.1 Solving nonlinear problems — 174
A.2 A code for the Newton–Raphson technique — 176

A.3 Obtaining the residuals and derivatives directly from virtual work	177
A.4 A linear solver	179
A.5 Coding the virtual work	181
A.6 Coding the equilibrium and constraint equations directly	184
A.7 Putting it all together: a program for static analysis of skeletal muscle	187
A.8 Example: static deformation of a cat medial gastrocnemius muscle	195
A.9 A variant of the previous example	199
Appendix B: Topics in time-dependent modelling	**202**
B.1 The finite-difference method	202
B.2 Example: dynamics of a one-degree-of-freedom system by central differences	204
B.3 Time-dependent problems: implicit methods	206
B.4 Example: a two-degrees-of-freedom system	212
B.5 A program for time-dependent analysis of skeletal muscle	215
B.6 Example: time-dependent deformation of a cat medial gastrocnemius muscle	225
References	233
Index	235

Preface

This book was motivated by our perception that although much work has been done in modelling skeletal muscle, little of this work is summarized in a systematic way. It was further motivated by the observation that treatments on theoretical models rarely include information on the biology of muscle; and vice versa, biological works lack the corresponding theory. Here, we attempt to fill these gaps. Because of the vast field of research, it was necessary to restrict the scope of this book to 'basic models'. Here, the word 'basic' is meant to indicate 'commonly used' and 'as originally introduced', it should not be interpreted to mean 'trivial'. Therefore, we may have chosen not to discuss special adaptions or most recent developments of a model.

The first part of this book (Chapters 1–4) contains a biological introduction to skeletal muscles, a description of selected muscle properties, and a mathematical treatment of Hill and Huxley type models. The second part (Chapters 5–7 and Appendices A and B) describes the mechanical and mathematical treatment of complex muscle models and concludes with considerations on movement control. Depending on the background of the reader, some chapters may contain well known information while others may appear extremely difficult. This is the nature of the topic and the nature of the book.

The different models of skeletal muscle discussed in this book all have their advantages, limitations, and applications. It was not our intent to compare models and judge them. Rather, the objective was to expose the readers to a variety of models and illustrate how these models might be used.

It was our intent to describe the basic biology, physiology and mechanics of muscular contraction, and to introduce the reader to a rigorous mathematical and mechanical approach for developing muscle models. We believe that we have accomplished this goal, despite the restriction to 'basic' models of muscle. Furthermore, we wanted to fill an obvious gap in this area of research: the lack of a systematic summary of the most frequently used muscle models. It is our sincere hope that we have also accomplished this goal, and that this book might inspire students and scientists alike.

Introduction

The modern era of modelling of skeletal muscle can be said to have been inaugurated in 1938 with the appearance in print of Hill's equation

$$(P + a)(v + b) = constant$$

This formula, relating the force P exerted by a muscle with its constant speed of shortening v, has retained its centrality in muscle modelling over the decades. It owes its importance to a number of reasons, not the least of which is that it was the result of careful experiments, which can be (and have been) reproduced in properly equipped laboratories. A second reason for the historical importance of this force–velocity relation is that an attempt was made by Hill to explain it in terms of a possible underlying physical mechanism; interestingly, both the experiments and the explanations are not purely mechanical, but include also a thermodynamic component. Finally, the resulting equation is mathematically simple, including only two constant parameters, a and b. As is the case with most scientific discoveries, the passage of time has revealed a few possible limitations, the most important of which being that the formula applies only to the experimental conditions under which it was obtained, namely, a maximally activated muscle under isometric conditions suddenly subjected to an isotonic release at a lower force level. A similar release at a higher than isometric force level does not abide by a similar formula. More importantly though, in more realistic situations, namely under simultaneous changes of neural activation, length and velocity, the formula cannot be applied directly. As far as the underlying physics is concerned, the discoveries concerning the microscopic structure of muscle, and the emergence of Huxley's cross-bridge model (1957), revealed a much more complex mechanism of contraction than Hill could have anticipated. Finally, the thermodynamic arguments which had originally led Hill to the derivation of his equation turned out not to be completely convincing.

Today, the researcher stands perplexed at the variety of phenomena displayed by skeletal muscle. These phenomena are the result of a complex interplay of chemical, electrical, and mechanical events at a variety of levels: macroscopic, cellular, molecular. The hope for a simple formula that will encompass all possible situations must, perhaps, be abandoned in favour of a number of soundly based complex models made of building blocks which take into consideration only those aspects relevant to a particular class of applications. Thus, for example, the question of filament compliance is probably of little relevance for most clinical and sports applications. This selectivity of the features to be included in any particular model, although somewhat disappointing at first sight, is actually the rule in the modelling of inert materials. In structural engineering applications, for example, the granular structure of steel is not directly incorporated in the material description, and the massive motion of dislocations which eventually takes place is only accounted for by macroscopic theories of plasticity. For a material scientist attempting to establish the cause of failure of a machine component, however, those neglected aspects may become of paramount importance.

As its title indicates, this book is neither a comprehensive treatise nor an exhaustive survey of the state of the art in the study of skeletal muscle. Its aim is to present in as complete and rigorous a manner as possible the basic physiological and mechanical concepts necessary to comprehend the foundations of a vast and active field of research. The first four chapters may well be used as the basis for a senior undergraduate course, while the last three chapters, more specialized in nature, are better suited for graduate students and researchers. What has been left out is much more extensive than what has been retained. Thus, the whole area of energetics and thermochemistry has not been visited, it being recognized that the excellent book by Woledge *et al.* [4] can be used as a supplementary text to cover those important topics. Even in the realm of mechanics proper, only straight-line models have been employed as a venue to illustrate how the rather powerful tools of mechanics, such as the principle of virtual work, are to be applied and numerically implemented in muscle models that: (a) can undergo arbitrarily large displacements and rotations; (b) are made of fibres which abide by very general constitutive laws; and (c) are subjected to general geometric constraints. The same basic principles and numerical techniques, however, are also applicable to other models, such as those based more directly on continuum mechanics and shell theory, through finite-element discretization [1]. A comprehensive survey of theoretical and experimental aspects pertaining to the musculo-skeletal system as a whole may be found in [3].

Assuming no previous familiarity with the anatomy and physiology of skeletal muscle, Chapter 1 introduces the reader to the fundamental concepts and terminology of the discipline, including some notions about the energetics and mechanism of force production. The stage thus set for more detailed considerations, Chapter 2 discusses the experimentally obtained force–length and force–velocity properties of muscle, and how they might be used in producing comprehensive models. Due attention is given to the inherent limitations of such models in that, in general, they do not account for history-dependent phenomena. The influence of passive elements on the overall response of the muscle–tendon complex is also addressed in this chapter. The next two chapters are devoted to the two pillars upon which most skeletal muscle models are established: the Hill (or rheological) and the Huxley (or structural) paradigms. While Chapter 3 deals

Introduction

with the conceptual biological framework wherein those two models are formulated, Chapter 4 tackles some of the mathematical issues associated with their implementation. This completes the first part of the book.

The second part of the book is mainly devoted to showing how the considerable knowledge acquired in the first four chapters may be used to produce a comprehensive muscle model which takes into consideration the complex geometry and architecture of a real muscle. In our experience, many such high-level models lack the rigour demanded by the intricacies of the fibre–aponeurosis–tendon assembly, which behaves as a highly nonlinear constrained system. For this reason, an entire chapter is devoted to a detailed presentation of the mechanics of deformable systems, with particular emphasis on the principle of virtual work. No previous familiarity with these topics is assumed on the part of the reader, so that Chapter 5 may be construed as a self-contained introduction to mechanics for the use of biomechanicians. By restricting the treatment to systems made up of line elements, the presentation is kept as simple as possible, given the inherent difficulties of the subject. The non-trivial role played by the geometric constraints is emphasized throughout. Based on the theory presented in Chapter 5, Chapter 6 develops complete computer codes for the static and dynamic analysis of a muscle–tendon complex. All the issues pertaining to the numerical implementation of the theory are presented in great detail in two appendices, which lead the reader in a step-by-step fashion through the actual computer routines. These appendices in themselves constitute a self-contained introduction to some of the fundamental numerical techniques for the analysis of complex systems: among the topics discussed therein are Gauss elimination for linear systems, Newton–Raphson approximation for nonlinear systems, and explicit and implicit methods for time-dependent behaviour. It is our hope that this type of presentation will serve to widen the horizons of young researchers and open their eyes to the immense field of possibilities of modelling afforded both by theoretical sophistication and by present and future computer power.

In spite of such theoretical and numerical sophistication, we are still far from a truly comprehensive model which will encompass all three main aspects of skeletal muscle, namely: (1) a rigorous formulation of the geometry and mechanics of the fibre–aponeurosis–tendon complex; (2) a thermochemical account of the energetics and its interaction with the mechanical events; and (3) a model of the control mechanism, whereby individual fibres and entire synergistic muscles are recruited and/or deactivated in response to mechanical and chemical stimuli. Chapter 7 addresses some of the issues pertaining to this last aspect. After a brief introduction to the neurophysiology of movement control, attention is focused on the problem of force sharing between synergistic muscles, whereby a redundant system of equations is encountered and solved by optimization techniques. In so doing, however, many other important aspects of movement control are touched upon that offer a panorama of this perhaps insufficiently explored subject.

The day of a comprehensive muscle model which will do justice to all three aspects (mechanical, thermochemical, and control) of the science of skeletal muscle is still ahead of us. We sincerely hope that this book will in some measure help the new generation of researchers to bring that day closer.

Our heartfelt thanks are due to many people who in one way or another helped to make this book a reality. Computer codes based on principles similar to some of those presented in Appendices A and B were developed at the University of Calgary in 1991 by then undergraduate engineering students Lisa Dumont and Ramesh Narayanasamy, under joint supervision with Prof. J.A. Hoffer, then at the University of Calgary Faculty of Medicine. Subsequently, in his MSc thesis [2] completed in 1993 at the Department of Mechanical Engineering, Johnny Mattar redeveloped and extended those codes to include the dynamic response. Mario Forcinito, now a doctoral candidate in the same department, has been a constant source of intellectual stimulation and exchange of ideas. Many students, researchers and professors associated with the Human Performance Laboratory at the University of Calgary Faculty of Kinesiology have contributed to the improvement of various sections of the book. Particular thanks are due to Holly Hanna, Tim Leonard, Ming-Ming Liu, Dale Oldham, and to Drs Todd Allinger, Antonio Guimarães, Tim Koh, Marco Vaz and John Wu. Individual chapters of the manuscript were kindly reviewed by Drs Jim Andrews, Jacques Bobet, Karin Gerritsen, Jim Hay, Brian MacIntosh, Benno Nigg, David Smith, Esther Suter and Ton van den Bogert, all of whom have contributed to the enhancement of the presentation and, in some cases, to the correction of mistakes. Naturally, all remaining mistakes are of our own doing. Finally, we acknowledge the support of the Natural Sciences and Engineering Research Council and of the Killam Foundation.

REFERENCES

[1] Anton, M.G. (1991) Mechanical models of skeletal muscle. PhD thesis, Department of Mechanical Engineering, The University of Calgary.
[2] Mattar, J. (1993) A dynamic model of skeletal muscle. MSc thesis, Department of Mechanical Engineering, The University of Calgary.
[3] Nigg, B.M. and Herzog, W., (eds) (1994) *Biomechanics of the Musculo-skeletal System*, John Wiley & Sons, Chichester.
[4] Woledge, R.C., Curtin, N.A. and Homsher, E. (1985) *Energetic Aspects of Muscle Contraction*, Academic Press, London.

Part One
Foundations

1
Basic introduction to skeletal muscle

1.1 GENERAL CONSIDERATIONS

Skeletal muscle is an organ that usually connects two bones. The organ 'muscle' consists primarily of skeletal muscle, but connective tissues internal and external to the muscle as well as nerve tissue play an integral part in the formation and function of muscle.

The primary function of muscle is to produce force and movements; however, muscle also produces a significant amount of heat upon contraction which can be used to regulate body temperature. When stimulated by a nerve, muscles shorten, provided they can overcome the external resistance imposed on them. Shortening and force production of muscle is referred to as 'contraction'.

There are three basic types of muscle: skeletal, cardiac, and smooth. In this book, we deal exclusively with skeletal muscle. Skeletal muscle, in contrast to cardiac and smooth muscles, can be controlled voluntarily. During contraction, skeletal muscles can produce force at a tremendous rate, and they can shorten at great speed and over large distances; thus they are extremely powerful. Muscles can also be controlled very accurately for precision tasks, such as handwriting.

1.2 MUSCLE STRUCTURE

Skeletal muscles are structurally organized in an intricate way, cross-sectionally and longitudinally. The entire muscle is surrounded by a layer of connective tissue called **fascia** and by a further connective tissue sheath known as the **epimysium** (Figure 1.1). The next smaller structure is the muscle bundle (**fascicle**), which consists of a number of muscle fibres surrounded by a connective tissue sheath called the **perimysium**. Then comes the **muscle fibre**, an individual muscle cell surrounded by a thin sheath of connective tissue (**endomysium**) which connects the individual fibres within a fascicle. Muscle fibres are cells with a delicate membrane, called the **sarcolemma**. Muscle fibres are made up of **myofibrils** (discrete bundles of myofilaments) lying parallel to one another. The systematic arrangement of the myofibrils gives muscle its typical striated pattern

4   *Theoretical models of skeletal muscle*

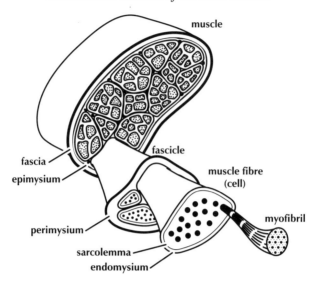

Figure 1.1 Schematic illustration of different structures and substructures of a muscle and a muscle fibre.

which is visible under a light microscope. The repeat unit in this pattern is a **sarcomere** (Figure 1.2). Sarcomeres are the basic contractile units of skeletal muscle; they are bordered by **Z-lines** (*Zwischenscheiben*). Z-lines are thin strands of protein extending perpendicular to the long axis of the myofibrils. Sarcomeres contain thick (**myosin**) and thin (**actin**) filaments, which are primarily made up of the protein molecules that give them their names. The Z-lines intersect the thin myofilaments at regular intervals.

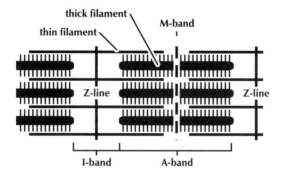

Figure 1.2 Schematic illustration of the basic contractile unit of the muscle, the sarcomere. (Adapted from [18].)

1.2.1 The thick filament

Thick filaments are typically located in the centre of the sarcomere. They cause the dark band of the striation pattern in skeletal muscles, referred to as the A- (anisotropic) band (Figure 1.2). A thick filament is made up of approximately 180 myosin molecules. Each myosin molecule has a molecular weight of 500 kDa, and contains a long tail portion consisting primarily of light meromyosin and a globular head portion, consisting of heavy meromyosin. The heads extend outward from the thick filament in pairs (Figure. 1.3). They contain a binding site for actin and an enzymatic site that catalyses the hydrolysis of adenosinetriphosphate (ATP) which releases the energy required for muscular contraction. Since the myosin heads have the ability to establish a link between the thick and thin filaments, they have been termed **cross-bridges**.

The myosin molecules in each half of the thick filament are arranged with their tail ends directed towards the centre of the filament. Therefore, the head portions are oriented in opposite directions for the two halves of the filament, and upon contraction (i.e. when myosin heads – cross-bridges – attach to the thin filament) the myosin heads pull the actin filaments towards the centre of the sarcomere.

The cross-bridges on the thick filament are about 14.3 nm apart longitudinally and are offset from each other by a 60° rotation (Figure 1.4). Since two cross-bridge pairs are offset by 180°, cross-bridge pairs with identical orientation are approximately 43 nm apart (3 × 14.3 nm).

1.2.2 The thin filament

The Z-lines bisect the thin filaments (Figure 1.2). Thin filaments appear light in the striation pattern, and the light band formed between the opposite ends of two thick filaments is called the I- (isotropic) band. The backbone of the thin filament consists of two helically interwoven chains of actin globules (Figure 1.5), whose diameter is about 5–6 nm. Thin filaments also contain the proteins tropomyosin and troponin. Tropomyosin is a long fibrous protein that lies in the grooves formed by the actin chains (Figure 1.5). Troponin is located at intervals of approximately 38.5 nm along the thin filament. Troponin is composed of three

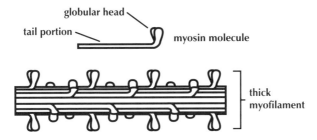

Figure 1.3 Schematic illustration of the thick myofilament. (From [20].)

6   *Theoretical models of skeletal muscle*

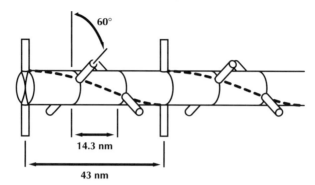

Figure 1.4 Schematic illustration of the arrangement of the cross-bridges on the thick filament. (From [18].)

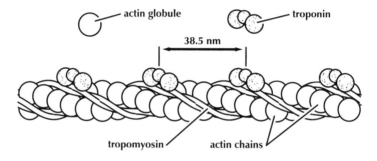

Figure 1.5 Schematic illustration of the thin myofilament, consisting of two helically interwoven chains of actin globules, tropomyosin, and troponin. (From [20].)

subunits: troponin C, which contains sites for calcium ion (Ca^{2+}) binding; troponin T, which contacts tropomyosin; and troponin I, which is thought to physically block the cross-bridge attachment site in the resting state (i.e. in the absence of Ca^{2+}).

When performing a cross-sectional cut through the zone of overlap between the thick and thin filaments in the sarcomere, it is revealed that each thick filament is surrounded by six thin filaments in a perfect hexagon (Figure 1.6). The cross-sections of the thick and thin filaments are approximately 12 and 6 nm in diameter, respectively. The distance between adjacent thick filaments is approximately 42 nm.

1.2.3 Other protein filaments in the sarcomere

Aside from the contractile proteins actin and myosin, skeletal muscle sarcomeres contain a variety of other proteins which are associated with structural and passive

Basic introduction to skeletal muscle 7

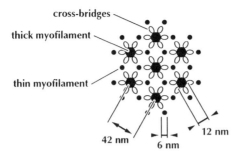

Figure 1.6 Schematic illustration of thick and thin myofilament arrangement in a cross-sectional view through the myofilament overlap zone.

functional properties of the sarcomere, rather than active force production. The most important of these proteins from a functional point of view is **titin**.

Titin is a huge (mass ≈ 3 MDa) protein which is found in abundance in myofibrils of vertebrate (and some invertebrate) striated muscle. Within the sarcomere, titin spans from the Z-line to the M-band, i.e. the centre of the thick filament. Although the exact functional role of titin remains to be elucidated, it is generally accepted that it acts as a molecular spring that develops tension when sarcomeres are stretched. Titin's location has prompted the idea that it might stabilize the thick filament within the centre of the sarcomere (Figure 1.7). Such stabilization may be necessary to prevent the thick filament from being pulled to one side of the sarcomere when the forces acting on each half of the thick filament are not exactly equal.

Evidence for the role of titin in thick filament centring has been provided by Horowits and colleagues [14–16] who showed that upon prolonged activation in chemically skinned rabbit psoas fibres, the thick myofilaments could easily be moved away from the centre of the sarcomere at short (< 2.5 µm) but not at long

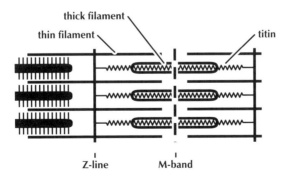

Figure 1.7 Schematic illustration of a sarcomere including titin proteins which are thought to run from Z-line to M-band, and, through their elastic properties, keep the thick myofilament centred in the sarcomere during contraction. (From [18].)

8 Theoretical models of skeletal muscle

(> 2.8 μm) sarcomere lengths when the titin 'spring' presumably was tensioned and so helped centre the thick myofilament.

1.2.4 Muscle shapes

The organizational structure of fibres, sarcomeres, and myofilaments is extraordinary, and so is the organizational structure of the fibres within a muscle. Skeletal muscles contain fibres from a few millimetres to several centimetres in length which are arranged either parallel to the longitudinal axis (i.e. parallel or **fusiform** muscles) or at a distinct angle to the longitudinal axis of the muscle (**pennate** muscles – Figure 1.8). Depending on the number (n) of distinct fibre directions, a muscle is called **unipennate** ($n = 1$), **bipennate** ($n = 2$), or **multipennate** ($n \geq 3$).

The variety of muscle shapes indicates the variety of functional tasks which need to be satisfied within an agonistic group of muscles. For example, the primary ankle extensor muscles in the cat are the two heads of the gastrocnemius, the soleus, and the plantaris. The soleus is essentially parallel-fibred. The medial head of the gastrocnemius and the plantaris are excellent examples of unipennate fibre arrangements, whereas the lateral head of the gastrocnemius is multipennate. The pennate muscles, by design, have shorter fibres than the parallel-fibred soleus [19]. Therefore, their absolute length range over which force can be generated (which is directly related to fibre lengths) is smaller compared to soleus. This arrangement makes perfect sense, considering that the soleus is a one-joint muscle which should ideally accommodate the entire range of ankle motion. The gastrocnemius and the plantaris cross the ankle and the knee, and during locomotion, flexion and extension of these two joints occur in such a way that a shortening of these muscles at one joint is (at least partly) offset by an elongation at the other joint, therefore requiring less excursion than the one-joint soleus.

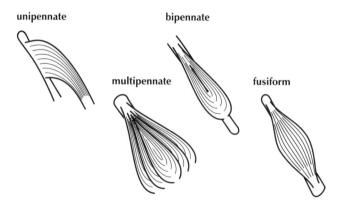

Figure 1.8 Classification of muscles into fusiform, unipennate, bipennate and multipennate.

1.3 MUSCLE FORCE CONTROL

1.3.1 Motor unit

Muscles receive the commands for force production from nerves. A single muscle nerve contains **afferent** and **efferent** axons. The afferent axons deliver information about the contractile status of the muscle to the central nervous system; the efferent axons deliver signals for contraction from the central nervous system to the muscle. The primary efferent pathways are called α motor neurones. Each α motor neurone innervates a number of muscle fibres. This functional unit (one α motor neurone and all the muscle fibres it innervates) is called a **motor unit**. A motor unit is the smallest control unit of a muscle, because all fibres belonging to the same motor neurone will always contract and relax in a synchronized manner.

The force in a muscle can be increased in two ways: (1) by increasing the number of active motor units; or (2) by increasing the frequency of stimulation to a motor unit. Since motor units are arranged in parallel (at least to a first approximation), the forces produced by individual motor units sum algebraically to the total muscle force. Therefore, the level of force can be controlled by activating or deactivating motor units.

When a motor unit receives a single stimulation pulse from its motor neurone, the corresponding force response is a single twitch. The average twitch duration is approximately 200 ms in a purely slow-twitch fibred muscle, such as the cat soleus (Figure 1.9). Therefore, if the soleus receives less than about five equally spaced stimulation pulses per second, it will show a series of individual twitch responses. When the stimulation frequency exceeds about 5 Hz, a second pulse will stimulate the muscle before the force effects of the first pulse have completely subsided. In this case, the force begins to accumulate, and with increasing frequencies, the force response becomes larger in magnitude and smoother (Figure 1.9). Therefore, aside from changing the number of active motor units, a muscle can adjust its force production by the frequency of stimulation of the motor units. For relatively low frequencies of motor unit stimulation (less than about 20 Hz for slow motor units, and less than about 50 Hz for fast motor units), there will be some force relaxation between stimulation pulses. The corresponding force–time history of such contractions has force 'ripples' (Figure 1.9, traces 6 (Hz), 10 (Hz), and 12.5 (Hz)). Such contractions are called **unfused tetanic** contractions. With increasing stimulation frequencies, the corresponding force–time traces become smoother until the force oscillations disappear; such contractions are referred to as **fused tetanic** contractions.

Sometimes contractions are referred to as submaximal, maximal, or supramaximal. During voluntary contractions, a **maximal** contraction corresponds to a maximal voluntary effort, and a **submaximal** contraction is any contraction which is less than maximal. During artificially elicited contractions of muscle – for example, by stimulating a muscle nerve using a stimulation electrode – force

10   *Theoretical models of skeletal muscle*

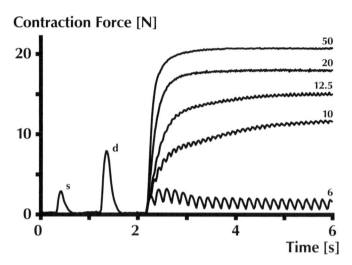

Figure 1.9 Single twitch (s), doublet twitch (d, two twitches separated by a 10 ms interval), and unfused and fused contractions of a cat soleus muscle at stimulation frequencies 6, 10, 12.5, 20, 50 Hz.

production may exceed that which a subject may produce voluntarily. Such artificial contractions are called **supramaximal**. They may be achieved, for example, by employing a stimulation frequency of all motor units (e.g. 100–150 Hz) which cannot be achieved by voluntary effort.

Motor units are composed of muscle fibres with similar biochemical and twitch properties. According to these properties, motor units are classified as fast or slow (there are more fibre types than just fast and slow, and the difference in the properties varies not discretely but in a continuous manner; however, these detailed differences are of no importance in the present context). Fast motor units, as the name implies, have a high maximal speed of shortening and a brief twitch duration. Slow motor units have a slower maximal speed of shortening and an increased twitch duration compared to the fast motor units. Furthermore, fast motor units are better equipped than slow motor units to produce muscle forces via anaerobic pathways (not involving oxygen), whereas slow motor units are better suited than fast motor units to produce the energy for muscular contraction via aerobic pathways (involving oxygen). Slow motor units are typically much more fatigue-resistant than fast motor units.

Morphologically, slow motor units are innervated by small-diameter motor neurones and they contain fewer muscle fibres than the corresponding fast motor units. This structural arrangement by itself is not particularly fascinating, except when viewed in the context of muscle force control. Henneman *et al.* [10] and Henneman and Olson [9] revealed the significance of the structural differences in slow and fast motor units. They showed that during a graded increase of force in a muscle, the small motor neurones innervating the small and slow motor units were recruited first. With increasing force demands, larger motor neurones innervating progressively larger motor units with increasingly fast-type properties were recruited. Therefore, a graded increase in force was accomplished by recruiting the smallest and slowest motor units first and the largest and fastest motor units last. This pattern of motor unit recruitment means that over a period of time there is a greater dependence on the slow motor units; the fast motor units are only recruited when particularly high forces are required. Since the small motor units typically have a large aerobic capacity, and thus, great endurance properties, and the fast motor units typically have little aerobic capacity, and thus fatigue quickly, the order of recruitment of motor units makes perfect sense; it is referred to as the **size principle** of motor unit recruitment.

The size principle was formulated and tested for the recruitment order of motor units within a given muscle. It is interesting to observe that there exist vast differences in the fibre type distribution of skeletal muscles, even within the same functional group. Again, the cat ankle extensors are a perfect example. The cat soleus is composed of primarily slow motor units (95–100%, [3,5]), whereas the medial gastrocnemius contains predominantly fast motor units (70–80% [3,6]). During quiet standing, it has been observed that the soleus produces substantial forces, while the medial gastrocnemius may be silent [12]. With increasing speeds of locomotion, the peak soleus forces remain roughly constant and the medial gastrocnemius forces increase several times [22]. Finally, during a rapid paw shake action (a movement which occurs at a frequency of about 10 Hz), soleus forces are low (or even zero), whereas the medial gastrocnemius activity is high [2,21]. This series of experiments illustrates the change in the functional role of the soleus and medial gastrocnemius for a variety of different movements. It is likely that the changes in the functional roles are tightly associated with the distribution of fibre types in these two muscles.

1.3.2 Excitation–contraction coupling

The process of excitation–contraction coupling involves the transmission of signals along nerve fibres, across the neuromuscular junction (where the end of the nerve meets the muscle fibre; Figure 1.10), and along muscle fibres. At rest, nerve and muscle fibres maintain a negative charge inside the cell compared to the outside (i.e. the membrane is polarized). Nerve and muscle fibres are excitable, which means that they can change the local membrane potential in a characteristic manner when stimuli exceed a certain threshold. When a muscle membrane becomes depolarized beyond a certain threshold, there is a sudden change in membrane permeability, particularly to positively charged sodium (Na^+) ions whose concentration outside the cell is much higher than that inside the cell. The resulting influx of Na^+ ions causes the charge inside the cell to become more positive. The membrane then decreases permeability to sodium and increases permeability to potassium ions which are maintained at a much higher concentration inside than outside the cell. The resulting outflow of the

12   *Theoretical models of skeletal muscle*

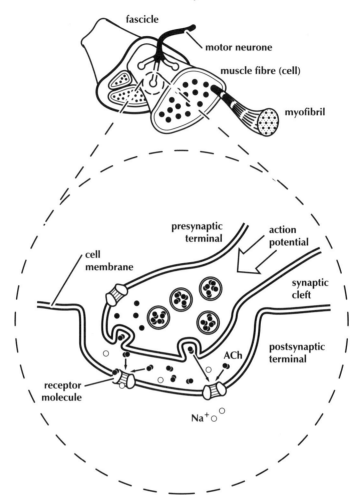

Figure 1.10 Schematic illustration of the neuromuscular junction with motor neurone and muscle cell membrane.

positively charged potassium ions causes a restoration of the polarized state of the excitable membrane. This transient change in membrane potential is referred to as an **action potential** and lasts for approximately 1 ms. In the muscle fibre, this action potential propagates along the fibre at a speed of about 5–10 m/s (Figure 1.11). In the α motor neurone, action potentials propagate at speeds in proportion to the diameter of the neurone, with the largest neurones (in mammals) conducting at 120 m/s.

The neuromuscular junction (Figure 1.10) is formed by an enlarged nerve terminal known as the **presynaptic terminal** that is embedded in small

Basic introduction to skeletal muscle 13

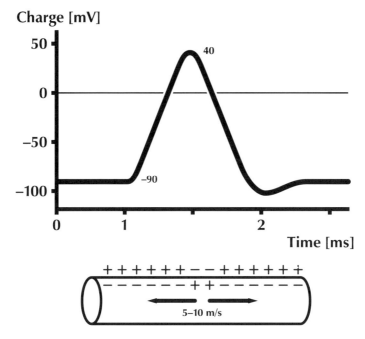

Figure 1.11 Schematic illustration of a single muscle fibre action potential (top) and the corresponding propagation of the action potential along the muscle fibre (bottom).

invaginations of the muscle cell membrane, the motor endplate, or **postsynaptic terminal**. The space between presynaptic and postsynaptic terminal is the **synaptic cleft**.

When an action potential of a motor neurone reaches the presynaptic terminal, a series of chemical reactions takes place that culminate in the release of acetylcholine (ACh) from synaptic vesicles located in the presynaptic terminal. ACh diffuses across the synaptic cleft, binds to receptor molecules of the membrane on the postsynaptic terminal, and causes an increase in permeability of the membrane to Na^+ ions. If the depolarization of the membrane due to Na^+ ion diffusion exceeds a critical threshold, then an action potential will be generated which travels along the stimulated muscle fibre. In order to prevent continuous stimulation of muscle fibres, ACh is rapidly broken down into acetic acid and choline by acetylcholinesterase which is liberally distributed in the postsynaptic membrane.

The action potential of the muscle fibre is not only propagated along and around the fibre, but also reaches the interior of the muscle fibre at invaginations of the cell membrane called T-tubules (Figure 1.12). Depolarization of the T-tubules causes the release of Ca^{2+} ions from the terminal cisternae of the sarcoplasmic reticulum (membranous sac-like structure which stores calcium) into the sarcoplasm surrounding the myofibrils. Ca^{2+} ions bind to specialized sites on the

14 Theoretical models of skeletal muscle

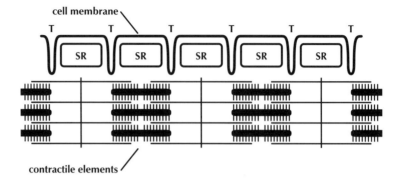

Figure 1.12 Schematic illustration of T-tubules (T) in a section of a muscle fibre and its association with the sarcoplasmic reticulum (SR) and the contractile myofilaments.

troponin molecules of the thin myofilaments, and so remove an inhibitory mechanism that otherwise prevents cross-bridge formations in the relaxed state (Figure 1.13). Cross-bridges then attach to the active sites of the thin filaments and, through the breakdown of ATP into adenosinediphosphate (ADP) plus a phosphate ion (P_i), the necessary energy is provided to cause the cross-bridge head to move and so attempt to pull the thin filaments past the thick filaments (Figure 1.14). At the end of the cross-bridge movement, an ATP molecule is

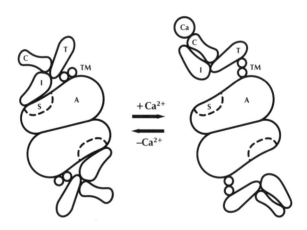

Figure 1.13 Schematic illustration of the inhibitory/excitatory regulation of cross-bridge attachment on the actin filament (A). Without calcium (left), the tropomyosin (TM) and troponin complex (troponin T, C, and I) are in a configuration which blocks the cross-bridge attachment site (S). Adding calcium (Ca^{2+}) to the calcium binding site of the troponin (troponin C) changes the configuration of the tropomyosin–troponin complex in such a way that the cross-bridge attachment site is exposed and cross-bridge attachment is possible.

Basic introduction to skeletal muscle 15

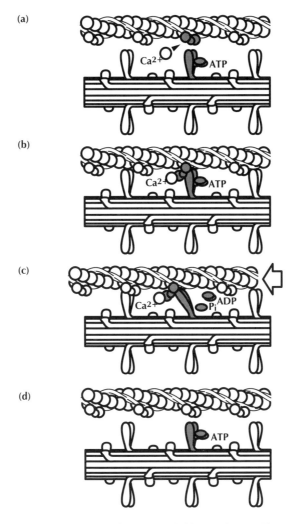

Figure 1.14 Schematic illustration of the cross-bridge cycle. (a) The muscle is at rest. The attachment site on the thin filament is covered by the tropomyosin–troponin complex. ATP is bound to the myosin cross-bridge. (b) Upon activation, calcium concentration increases in the sarcoplasm and calcium (Ca^{2+}) binds to troponin C, thereby causing a configurational change which exposes the actin binding site. (c) The cross-bridge attaches to the actin and goes through a configurational change. The splitting of ATP into ADP and P_i provides the energy which results in contraction, i.e. movement of the thin past the thick filaments. (d) A new ATP attaches to the cross-bridge and the cross-bridge can detach from the thin filament and is ready for a new interaction with (another) attachment site on the thin filament.

thought to attach to the myosin portion of the cross-bridge so that the cross-bridge can release from its attachment site, go back to its original configuration, and be ready for a new cycle of attachment. This cycle repeats itself as long as the muscle fibre is stimulated. When stimulation stops, Ca^{2+} ions are actively transported back into the sarcoplasmic reticulum, resulting in a decrease of Ca^{2+} ions in the sarcoplasm. As a consequence, Ca^{2+} ions diffuse away from the binding sites on the troponin molecule, and cross-bridge cycling stops.

1.4 ENERGY CONSIDERATIONS

The physiology of muscular work is a process of transforming chemical energy into mechanical energy. Chemical reactions involving transfer of energy either liberate energy (**exergonic reactions**) which, in skeletal muscle may be used to produce work, or require energy (**endergonic reactions**). In skeletal muscle (as in many other cells), ATP functions as an energy storage device. ATP is a molecule in which much of the total energy content of the compound is concentrated in the phosphate bonds. Therefore, cleavage of a phosphate from ATP (in a hydrolytic reaction) is an exergonic process which yields energy:

$$ATP^{4-} + H_2O \rightarrow ADP^{3-} + P_i^{2-} + H^+ + energy \qquad (1.1)$$

It is difficult to measure the energy liberated by hydrolysing one mole of ATP. The reason for this difficulty lies in the fact that the reactants and products of equation (1.1) inside the muscle fibre are not known precisely; they may fluctuate or may be compartmentalized. Furthermore, the energy yield depends on the temperature and the pH value inside the cell, both of which are known to change during muscular work. Values which are typically associated with the *in vivo* energy yield for the above equation range from -40 to -50 kJ/mol.

The ATP stores in a muscle are limited, therefore ATP needs to be continuously replenished during muscular work. Phosphocreatine (PCr) is a high-energy phosphate compound which functions as an immediate source for ATP regeneration in skeletal muscle. PCr combines with ADP to form ATP:

$$H^+ + PCr^{2-} + MgADP^- \rightarrow Cr + MgATP^{2-} \qquad (1.2)$$

This reaction, which has been shown to occur at a similar rate to the ATP hydrolysis in normally functioning muscle, is endergonic and requires about 14 kJ/mol to proceed in the direction of the arrow shown in the above equation.

For contractions of short duration, it has been assumed that ATP hydrolysis and ATP resynthesis with the help of PCr are the primary chemical processes. Since ATP hydrolysis is an exergonic reaction (-40 to -50 kJ/mol) and ATP resynthesis is an endergonic reaction (14 kJ/mol), the net energy of the combined reactions which is available for muscular work is -26 to -36 kJ/mol. The best single value available at present for the net energy yield of these two reactions is probably -34 kJ/mol [23]. Phosphocreatine is a limited source for ATP resynthesis. During strenuous muscular work, PCr stores are depleted within

Basic introduction to skeletal muscle 17

seconds. Therefore, for extended periods of work, ATP must be resynthesized in other ways. The energy source for this continued process comes from the breakdown of complex molecules (food) to simple molecules. The main stores for ATP resynthesis are carbohydrates in the form of glycogen and fatty acids in the form of triglycerides.

Carbohydrates may be broken down aerobically or anaerobically, whereas fatty acids can only be broken down aerobically. The initial stage of carbohydrate breakdown to pyruvate is anaerobic; it occurs in the cytoplasm of the muscle fibre. Without the use of oxygen, the glycogen breakdown will give a net yield of three ATP molecules according to the following relation:

$$\text{glycogen unit} + 3\text{MgADP}^- + 3\text{HPO}_4^{2-} + \text{H}^+ \rightarrow 2 \text{ lactate} + 3\text{MgATP}^{2-} \quad (1.3)$$

Glucose breakdown yields one less ATP molecule than glycogen breakdown because an extra ATP is required in the early stages of glycolysis. The by-product of the anaerobic ATP resynthesis is lactic acid. Lactic acid is thought to contribute to muscle fatigue.

In the aerobic scheme of carbohydrate breakdown, the initial step to arrive at pyruvate from glycogen or glucose is the same as in the anaerobic pathway. If there is sufficient oxygen, little or none of the pyruvate will be reduced to lactate; rather pyruvate will enter the mitochondria (specialized organelles in muscle fibres where the aerobic oxidation of carbohydrate and fatty acids occurs). The aerobic mitochondrial oxidation of pyruvate results in 34 ATPs; therefore, together with the two or three ATPs obtained in the initial anaerobic pathway, the total aerobic yield of a glycogen/glucose monomer breakdown is 36–37 ATPs:

$$\text{glycogen unit} + 6\text{O}_2 + 36\text{MgADP}^- + 36\text{HPO}_4^{2-} + 36\text{H}^+ \rightarrow$$
$$6\text{CO}_2 + 6\text{H}_2\text{O} + 36\text{MgATP}^{2-} \quad (1.4)$$

Fatty acids are oxidized inside the mitochondria in a purely aerobic way. The average yield from one mole of a mixture of fatty acids is about 138 moles of ATP. For palmitate, an abundant fatty acid in the human body, the energy balance is as follows [4]:

$$\text{C}_{15}\text{H}_{31}\text{COOH} + 129\text{P}_i + 129\text{ADP} + 23\text{O}_2 \rightarrow 12\text{CO}_2 + 16\text{H}_2\text{O} + 129\text{ATP} \quad (1.5)$$

In a maximally working muscle, ATP and PCr stores are depleted within a few seconds. For example, it is well known that track and field athletes who can sprint a 100 m distance in about 10 s are unable to maintain maximal sprinting speed for the entire duration of the race. The slowing down towards the end of the 10 s race is likely to be associated with a depletion of the ATP and PCr stores.

When extended periods of muscular work are required, the power output must be reduced from that of a sprint-type effort so that ATP and PCr stores can be maintained at approximately constant levels through the aerobic breakdown of glycogen and fatty acids.

One of the interests in determining the net energy yield from biochemical reactions within muscle has been the study of energy balance during muscular contraction. The aim of energy balance studies is to understand the workings of muscle in a better way by comparing the energy produced during muscle contraction with the amount of energy that is made available by the chemical reactions thought to take place during contraction. If the exact extent of each reaction as well as the corresponding energy yield (or energy requirement), i.e. the change in molar enthalpy, were known, then the net energy yield of all contractions should equal the energy produced by the muscle in the form of heat and mechanical work.

Much work on the energy balance of muscular contraction has been performed in amphibian skeletal muscle. In these experiments, it is typically assumed that the ATP hydrolysis reaction and the resynthesis of ATP through PCr are the principal reactions during short muscle contractions. Furthermore, it is typically assumed that heat and work production have to be accounted for immediately by the corresponding chemical reactions, i.e. there is little or no time delay between the chemical reactions and the heat and work production of the muscle. Using these assumptions (which appear quite feasible), the total energy yield of the chemical reactions (the explained energy) is usually considerably less (typically about 30%) than the energy measured in the form of heat and work (the observed energy). Obviously, one cannot obtain more energy during muscular contraction than the energy that is liberated by the sum of all chemical reactions. Therefore, there must be an error in the assumptions or the measurements. Assuming that heat and work can be measured accurately, the following factors might account for the discrepancy between the explained and the observed energy: not all chemical reactions are accounted for; the energy yield (requirement) for given reactions in the live muscle might not be represented adequately; or the energy balance between chemical reactions and heat and work production may not be immediate but time-delayed. Perfect energy balance measurements across a variety of muscles and contractile conditions have not been made to date, suggesting that there are still unknown factors which have not been accounted for in the energy balance experiments.

1.5 TYPES OF CONTRACTIONS

A stimulated muscle produces force and will shorten, provided that shortening is not prevented by some external constraints. A shortening contraction is referred to as a **concentric** contraction. If the ends of the muscle (or an isolated fibre or sarcomere) are fixed rigidly during a contraction, the contraction is called **isometric** (i.e. there is no change in muscle, fibre, or sarcomere length). Finally, if the external force exceeds the isometric force capability of a contracting muscle, the muscle (fibre, sarcomere) is forcibly stretched and the resulting contraction is called **eccentric**. It must be pointed out that the type of contraction may depend on the level at which it is observed. For example, it has been well documented that during an isometric contraction of an isolated fibre, sarcomeres

Basic introduction to skeletal muscle 19

within that fibre may contract concentrically (typically towards the end regions of the fibre) and eccentrically (in the middle portion of the fibre [8]). Furthermore, it has been demonstrated that muscle fibres of the cat medial gastrocnemius may shorten at the same time as the entire muscle is elongating [13].

During normal everyday movements, dynamic contractions probably occur at changing speeds and continuously varying force levels [11]. However, during experimental tests aimed at elucidating the mechanical properties of skeletal muscles, it has been found useful to control either the speed or the force of contraction. Movements performed at constant speeds (concentrically or eccentrically) are referred to as **isokinetic**; contractions performed at a constant resistive force are referred to as **isotonic**.

1.6 HOW DOES A MUSCLE PRODUCE FORCE?

The molecular mechanisms of muscular force production are explained by the cross-bridge theory. The basic features of the cross-bridge theory were introduced by Huxley [17], and although modifications of the 1957 cross-bridge theory [17] have been proposed, the basic principles of the early theory are still accepted by the majority of muscle physiologists. The detailed aspects of the cross-bridge theory are discussed later in this book; here a conceptual overview of the theory is given.

In the cross-bridge theory, it is assumed that muscle contraction occurs through the cyclic interaction of myosin cross-bridges with the actin filaments (Figure 1.14). The cross-bridges, which extend laterally from the thick filaments, are thought to attach to specialized binding sites on the thin filament. Movement and force production occur through the rotation of the cross-bridge heads, thereby pulling the thin filaments across the thick filaments towards the centre of the sarcomere. Sliding of the myofilaments relative to each other is assumed to occur with minimal deformation of the filaments (i.e. they are treated as rigid elements), and the energy for the contractile process is linked to the hydrolysis of ATP – one ATP molecule per cross-bridge cycle.

It should be pointed out that there exist alternative theories of force production [18]. However, at present, these theories are not generally accepted despite the fact that the cross-bridge theory does not explain all properties observed in skeletal muscle. Specifically, the dependence of force production on the history of contraction is not part of the cross-bridge model, although history-dependent phenomena of force production have been observed before the formulation of the initial cross-bridge paradigm [1]. Also, many energetic aspects of muscular contraction observed experimentally are not well explained in the cross-bridge formulation, particularly those associated with eccentric contractions. Nevertheless, at present, the cross-bridge theory explains more phenomena in a better way than any competing theory.

An interesting development regarding myofilament deformation has occurred in the past two years. occurred in the mid-1990s when it was found that thin (and possibly thick) myofilaments were compliant. In the traditional formulation of the cross-bridge theory, myofilaments are considered as strictly rigid elements. This treatment of the myofilaments has several important practical and theoretical implications. For example, if myofilaments are rigid and all the sarcomere elasticity is contained in the cross-bridges, stiffness measurements may be directly linked to the number of attached cross-bridges. In fact, assuming that cross-bridge stiffness is linear (as has typically been done [17]), sarcomere stiffness should be linearly proportional to the number of attached cross-bridges. For the theoretical formulation of the cross-bridge theory, rigid myofilaments imply that the relative speed of an arbitrary point on the thin filament relative to an arbitrary point on the thick filament is always given by the relative speed between the thin and thick filament [17]. For compliant filaments, the determination of the speed of a point on the thin relative to a point on the thick filament would have to consider the average relative speed of myofilament movement and the local relative speed caused by the myofilament compliance [7].

PROBLEMS

1.1 The size principle of motor unit recruitment suggests that during a graded contraction with increasing force, motor units are recruited according to size – small motor units first, large motor units last.

   a) What physiological advantages does a recruitment order according to the size principle have?

   b) How could one test the correctness of the size principle in an *in vivo* experiment on human skeletal muscle non-invasively (i.e. the skin is not penetrated)?

1.2 A few years ago, one of the basic questions related to skeletal muscle fatigue was whether loss of force during fatiguing exercise was caused by a loss of the central nervous drive to the muscle or by a loss of the contractile ability of the muscle. How could one test which of the two proposed mechanisms of muscle fatigue was correct during *in vivo* human, voluntary contractions?

1.3 At present, it is assumed in the cross-bridge theory of muscle contraction that each cross-bridge cycle requires the energy from hydrolysing one ATP molecule. Develop a conceptual experiment to test the hypothesis that exactly one ATP molecule is hydrolysed per cross-bridge cycle. All imaginable techniques are at your disposal.

1.4 In this chapter, the cat ankle extensor muscles were used to show the diversity in structure, fibre type distribution, and functional properties of muscles. Perform a literature search and attempt to find quantitative data for the corresponding human muscles (soleus, gastrocnemius, and plantaris), including their size (volume, weight), fibre type distribution, fibre length, fibre orientation, action during cyclic movements such as walking and running, and contractile properties (force–length and force–velocity properties).

REFERENCES

[1] Abbott, B.C. and Aubert, X.M. (1952) The force exerted by active striated muscle during and after change of length. *J. Physiol.* (London), **117**: 77–86.
[2] Abraham, L.D. and Loeb, G.E. (1985) The distal hindlimb musculature of the cat. *Exp. Brain Res.*, **58**: 580–593.
[3] Ariano, M.A., Armstrong, R.B. and Edgerton, V.R. (1973) Hindlimb muscle fiber populations of five mammals. *J. Histochem. Cytochem.*, **21(1)**: 51–55.
[4] Astrand, P.O. and Rodahl, K. (1977) *Textbook of Work Physiology*. McGraw-Hill, New York, pp. 11–34.
[5] Burke, R.E., Levine, D.N., Saleman, M. and Tsairis, P. (1974) Motor units in cat soleus muscle: physiological, histochemical and morphological characteristics. *J. Physiol.* (London), **238**: 503–514.
[6] Burke, R.E. and Tsairis, P. (1973) Anatomy and innervation ratios in motor units of cat gastrocnemius. *J. Physiol.* (London), **234**: 749–765.
[7] Forcinito, M., Epstein, M. and Herzog, W. (1997) Theoretical considerations on myofibril stiffness. *Biophys. J.*, **72**: 1278–1286.
[8] Gordon, A.M., Huxley, A.F. and Julian, F.J. (1966) The variation in isometric tension with sarcomere length in vertebrate muscle fibres. *J. Physiol.* (London), **184**: 170–192.
[9] Henneman, E. and Olson, C.B. (1965) Relations between structure and function in the design of skeletal muscles. *J. Neurophysiol.*, **28**: 581–598.
[10] Henneman, E., Somjen, G. and Carpenter, D.O. (1965) Functional significance of cell size in spinal motor neurones. *J. Neurophysiol.*, **28**: 560–580.
[11] Herzog, W., Leonard, T.R. and Guimarñães, A.C.S. (1993) Forces in gastrocnemius, soleus, and plantaris tendons of the freely moving cat. *J. Biomech.*, **26**: 945–953.
[12] Hodgson, J.A. (1983) The relationship between soleus and gastrocnemius muscle activity in conscious cats – a model for motor unit recruitment? *J. Physiol.* (London), **337**: 553–562.
[13] Hoffer, J.A., Caputi, A.A., Pose, I.E. and Griffiths. R.I. (1989) Roles of muscle activity and load on the relationship between muscle spindle length and whole muscle length in the freely walking cat, in *Progress in Brain Research.* (eds J.H.H. Allum and M. Hulliger), Elsevier Science Publishers, Amsterdam, pp. 75–85
[14] Horowits, R., Maruyama, K. and Podolsky, R.J. (1989) Elastic behaviour of connectin filaments during thick filament movement in activated skeletal muscle. *J. Cell Biol.* **109**: 2169–2176.
[15] Horowits, R. and Podolsky, R.J. (1987) The positional stability of thick filaments in activated skeletal muscle depends on sarcomere length: evidence for the role of titin filaments. *J. Cell Biol.*, **105**: 2217–2223.
[16] Horowits, R. and Podolsky, R.J. (1988) Thick filament movement and isometric tension in activated skeletal muscle. *Biophys. J.*, **54**: 165–171.
[17] Huxley, A.F. (1957) Muscle structure and theories of contraction. *Prog. Biophys. Biophys. Chem.*, **7**: 255–318.
[18] Pollack, G.H. (1990) *Muscles and Molecules: Uncovering the Principles of Biological Motion.* Ebner and Sons, Seattle.
[19] Sacks, R.D. and Roy, R.R. (1982) Architecture of the hind limb muscles of cats: Functional significance. *J. Morph.*, **173**: 185–195.

[20] Seeley, R.R., Stephens, T.D. and Tate, P. (1989) *Anatomy and Physiology*, Times Mirror/Mosby College Publishing, Toronto.
[21] Smith, J.L., Betts, B., Edgerton, V.R. and Zernicke, R.F. (1980) Rapid ankle extension during paw shakes: Selective recruitment of fast ankle extensors. *J. Neurophysiol.*, **43**: 612–620.
[22] Walmsley, B., Hodgson, J.A. and Burke, R.E. (1978) Forces produced by medial gastrocnemius and soleus muscles during locomotion in freely moving cats. *J. Neurophysiol.*, **41**: 1203–1215.
[23] Woledge, R.C., Curtin, N.A. and Homsher, E. (1985) *Energetic Aspects of Muscle Contraction*. Academic Press, London.

2
Modelling skeletal muscle using simple geometric shapes: biological considerations

2.1 INTRODUCTION

Skeletal muscles come in a variety of sizes and shapes, depending on the function they fulfil; and the arrangement of the contractile elements (fibres) within a muscle knows no limits. In many situations, it is imperative to know how the geometry of a muscle changes during contraction, because these changes may directly influence force production. For example, in a pennate muscle the long axis of a fibre (and thus, presumably the direction of force production) is at a distinct angle, α, with respect to the line of action of the muscle (Figure 2.1). Therefore, if one wants to know the force component of the fibre (f_c) in the direction of the line of action of the muscle, the angle α, usually referred to as the angle of pinnation, needs to be known at any instant in time, because

$$f_c = f \cos(\alpha) \qquad (2.1)$$

where f is the force produced by the fibre. In the simplest case, it may be assumed that the angle of pinnation is constant during contractions; however, there is

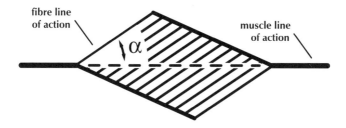

Figure 2.1 Definition of the angle of pinnation, α.

24 Theoretical models of skeletal muscle

ample evidence to the contrary [15, 51, 95, 100, 103]. A note of caution must be added here regarding equation (2.1). Although the component of the fibre force along the line of action of a muscle can be calculated this way, it is generally wrong to assume that the sum of all fibre components along the line of action is the total muscle force. We will come back to this point later in the book and will demonstrate that, because of the constraints imposed on the system 'muscle' (i.e. volume conservation), the resultant muscle force (as measured, for example, using a tendon transducer) cannot be divided into a fibre and an aponeurosis component using a three-force free-body diagram.

Another example in which the geometry of the muscle may affect force production indirectly is related to the muscle spindles. Muscle spindles are proprioceptors (i.e. sensors which detect length changes and changes in the speed of contraction in a muscle) embedded in parallel with the fibres in skeletal muscle [77]. Stretching of the muscle spindles typically causes excitation of the stretched muscle, and so an increase in force [51, 77]. Therefore, when studying the possible influence of spindle-mediated force production during movements, it is essential to know the instantaneous lengths of the muscle spindles. Often, spindle lengths have been derived directly from the muscle–tendon lengths; however, it has been demonstrated that muscle–tendon length is not necessarily a good indicator of muscle fibre and muscle spindle lengths [35, 36, 51].

Many models of skeletal muscle, particularly unipennate muscles, are geometric (e.g. [103, 104]). The geometric models of muscle discussed in this chapter use the geometry for the derivation of equations describing the contractile behaviour. Typically, these models are bound by straight-line segments so that analytic expressions for their contractile behaviour may be derived. Collectively, we will be referring to these models as straight-line models (SLMs).

Straight-line models always contain active force-producing elements (the contractile elements or muscle fibres), and they may contain passive elements. The passive elements may be in series or in parallel with the force-producing elements. The passive elements arranged in series are normally associated with the tendons or the aponeuroses of the muscle. Other passive, in-series elements, although they exist (e.g. elastic elements within the cross-bridges [52], or the myofilaments [70,90]), are typically not considered in SLMs. The passive elements arranged in parallel with the force-producing elements are typically related to the connective tissues surrounding muscle fibres, muscle bundles, and the entire muscle.

2.2 SHAPES

There are two basic shapes of SLMs of skeletal muscle: parallel-fibred models and pennate-fibred models (Figure 2.2). Parallel-fibred or fusiform muscles have the fibres aligned in parallel (or very close) to the line of action of the muscle; pennate muscles have the fibres aligned at a distinct angle (the angle of pinnation) relative to the line of action of the muscle. Pennate muscles are further

Modelling skeletal muscle using simple geometric shapes 25

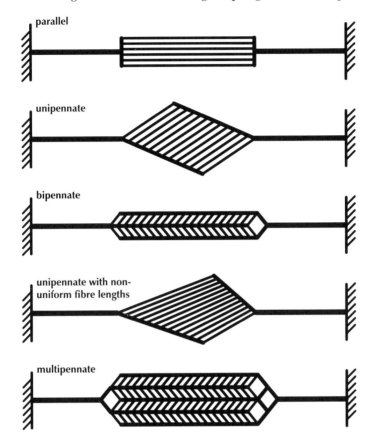

Figure 2.2 Schematic representation of different geometries found in mammalian skeletal muscles.

grouped according to the number of distinctly different fibre directions they contain. In a unipennate muscle all fibres are aligned in one direction; in a bipennate muscle the fibres are aligned in two distinctly different directions; and in a multipennate muscle, the number of distinct fibre directions exceeds two (Figure 2.2).

The shapes of straight-line models may further be changed by not restricting the fibre lengths to be uniform within a muscle (Figure 2.2). By changing the fibre direction and the length of the fibres, an infinite number of muscle shapes may be modelled.

2.3 CONTRACTILE ELEMENT PROPERTIES

Active force in SLMs is produced by the contractile elements. Therefore, the properties given to the contractile elements directly impact on the muscle's

26   *Theoretical models of skeletal muscle*

contractile behaviour. Most often, the contractile elements of SLMs contain force–length and force–velocity or power–velocity properties. These properties will be introduced in the following, together with the less frequently used history-dependent properties. However, before introducing these properties, it should be emphasized that the force–length, force–velocity, and the history-dependent properties have been explained (at least in part) by the mechanism of muscular force production (e.g. the cross-bridge model [52, 53]). Here, these explanations are omitted, because they are not required to understand SLMs, and because SLMs are not aimed at investigating the mechanisms of force production. Attempts to explain these properties will be made in Chapters 3 and 4, particularly when discussing the cross-bridge model.

2.3.1 Force–length properties of skeletal muscle

The force–length property of a muscle is defined by the maximal isometric force a muscle can exert as a function of its length. The fact that force production in skeletal muscle is length-dependent has been known for a long time [13]. Force–length properties have been derived for sarcomeres [32], for isolated fibres [91], and for entire muscles [33, 43, 46]. When discussing the force–length properties of SLMs of muscles, the fibre and muscle properties (rather than the properties of isolated sarcomeres) are of interest.

Force–length properties may be described in the simplest case by a symmetric curve and the corresponding peak force (F_0) and working range (L) [103]. An example of two force–length curves with different peak force values and working ranges is shown in Figure 2.3. Muscle 1 has a large cross-sectional area and corresponding peak force, F_{01}, and short fibres and a corresponding small working range, L_1; muscle 2 has a small peak force, F_{02}, and a large working range, L_2.

The maximal isometric force that a muscle can exert is primarily a function of the physiological cross-sectional area (PCSA):

$$PCSA = \frac{\text{Muscle Volume}}{\text{Fibre Length}} \quad (2.2)$$

where the denominator is the mean length of muscle fibres at optimal sarcomere lengths. (Optimal sarcomere length is the length at which a sarcomere can produce maximal force, e.g. about 2.1 µm, 2.4 µm, and 2.7 µm for frog, cat, and human skeletal muscles, respectively; [42,98]). The physiological cross-sectional areas, therefore, may be thought of as the sum of the cross-sectional area of all fibres arranged in parallel within a muscle. The maximal isometric force, F_0, is typically calculated from the known PCSA using a proportionality factor, i.e.

$$F_0 = k \cdot PCSA \quad (2.3)$$

Modelling skeletal muscle using simple geometric shapes

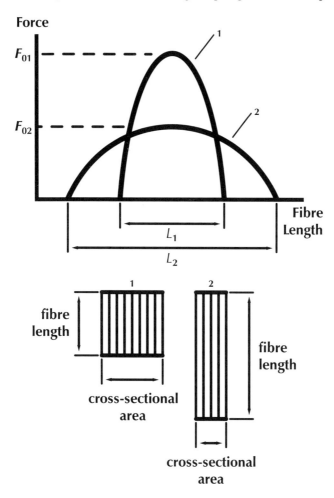

Figure 2.3 Schematic force–length relation of two muscles with different cross-sectional areas and fibre lengths.

where k is typically taken as 20–40 N/cm^2; however, k-values used in the literature range from 9.8 N/cm^2 [99] to 147 N/cm^2 [75].

The working range of the muscle, L, may be approximated by

$$L = \text{Optimal Fibre Length} \qquad (2.4)$$

indicating that the working range is approximately equal to the mean length of the muscle fibres at optimal sarcomere lengths. This relation between the working range and optimal fibre length makes sense when realizing that the working range of a sarcomere corresponds approximately to the optimal length

28 Theoretical models of skeletal muscle

of a sarcomere [32], and that fibres contain sarcomeres in series along their entire length.

The sarcomere force–length relation depends on the lengths of the thick and thin myofilaments. Once these lengths are known, the exact relation between sarcomere length and force can be calculated based on the cross-bridge theory [32, 42, 52]. For frog skeletal muscle, the equations relating to sarcomere lengths (SL) in micrometres and normalized force (F) (the force divided by the maximal isometric force) are described by four straight lines (Figure 2.4):

$$\begin{aligned}F &= -2.667 + 2.1\,SL & 1.27\,\mu m \leqslant SL &< 1.67\,\mu m \\ F &= 0.04 + 0.48\,SL & 1.67\,\mu m \leqslant SL &< 2.00\,\mu m \\ F &= 1.0 & 2.00\,\mu m \leqslant SL &< 2.25\,\mu m \\ F &= 1.592 - 0.71\,SL & 2.25\,\mu m \leqslant SL &< 3.65\,\mu m\end{aligned} \quad (2.5)$$

For entire muscles, the relation between normalized force (F) and muscle fibre lengths (L) has been approximated by (Figure 2.5)

$$F = -6.25(L/L_0)^2 + 12.5(L/L_0) - 5.25 \quad (2.6)$$

where L_0 is the optimal muscle length [103]. Note that equation (2.6) gives a working range, R, of the muscle of $0.8L_0$.

In the sarcomere force–length relationship shown in Figure 2.4, the region of positive slope (i.e. for sarcomere lengths ranging from 1.27 μm to 2.00 μm) is called the **ascending limb** of the force–length relationship; the area of zero slope (2.0 – 2.25 μm) is the **plateau region**; and the region of negative slope

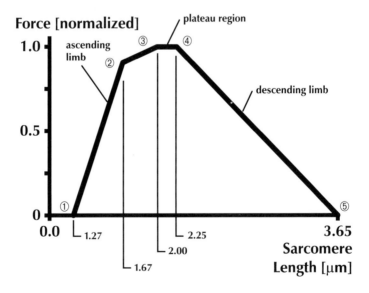

Figure 2.4 The sarcomere force–length relation of frog skeletal muscle.

Modelling skeletal muscle using simple geometric shapes

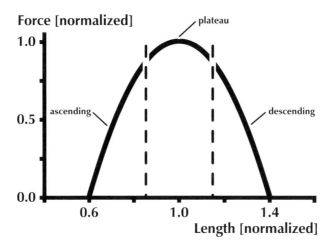

Figure 2.5 The normalized force–length relation of mammalian skeletal muscle obeying equation (2.6) [From 103].

(2.25–3.65 μm) is the **descending limb**. These regions have also been defined (albeit less rigorously) for the force–length relationships of entire muscles (Figure 2.5).

The mechanical behaviour of skeletal muscle and sarcomeres on the descending limb of the force–length relationship has been a topic of great debate. Almost half a century ago, it was proposed that sarcomere behaviour was unstable on the descending limb of the force–length relationship [49]. Instability was defined as a divergence of sarcomere length upon stimulation. Instability of sarcomere length was said to occur because, on the descending limb of the force–length relationship, a sarcomere that is shorter than its neighbouring sarcomere would also be stronger, and therefore, upon activation, the 'short, strong' sarcomere would shorten and the 'long, weak' sarcomere would be stretched, thereby creating an unstable sarcomere length behaviour. The idea of sarcomere instability on the descending limb of the force–length relationship is intuitively appealing for a variety of reasons. First, the negative slope of the descending limb has the characteristics of a 'softening' material, i.e. a material whose force becomes less as it is stretched. Softening materials are unstable, therefore viewing muscles on the descending limb as unstable is convenient and, at a superficial glance, appears to be correct. Secondly, experimental measurements of force on the descending limb of the force–length relationship exhibit a 'creep-like' behaviour, i.e. the force continues to increase after full activation has been reached [32]. Again, it is convenient to explain this force creep with sarcomere instability, particularly since similar phenomena are not observed when tests are performed on the ascending limb or plateau region of the force–length relationship. Thirdly, the amount of length change differs between sarcomeres within a fibre during isometric contractions on the descending limb of the force–length relationship. Typically, sarcomeres located towards the end regions of fibres shorten and sarcomeres located in the middle region of fibres are stretched in these situations (e.g. [56,60]).

In our opinion, the issue of instability of sarcomere length and force has caused confusion in the past. Implementing a force–length relationship of skeletal muscle as shown, for example, in Figure 2.4 will invariably lead to force instability on the descending part of the relationship. However, one must ask why should the basic contractile element of skeletal muscle be built in such a way that it is unstable over about 60% of its operating range, therefore leaving just 40% for safe use? And why is the known dynamic behaviour of sarcomeres ignored in favour of the static force–length relationship?

A system is said to be stable if, following a perturbation, the system will return to its original state. For skeletal muscles, sarcomere stability implies that any perturbation in sarcomere length will not cause a divergence of sarcomere length from some initial state. Sarcomeres arranged in series within a muscle are stable if the stiffness of all sarcomeres is positive [6]. The descending limb of the force–length relationship has a negative slope, and therefore, sarcomere stiffness would be negative if this relationship represented the true sarcomere properties. However, it is well known that the force–length relationship, particularly the descending part of it, does not represent dynamic properties of sarcomeres or muscle fibres, not even for very slow (quasi-static) stretching or shortening. Force–length relationships are obtained for isometric contractions at discrete length. The relationships, therefore, represent a series of isolated and independent static observations. If a sarcomere or muscle fibre was fully activated during an isometric contraction, and then was stretched or shortened to a new length, the force would not follow the 'continuous' descending limb shown in Figure 2.4. Rather, for a stretch, the force would be larger, and for shortening, the force would be lower than the corresponding theoretical force obtained from the isometric force–length relationship (Figure 2.6). This result indicates that assertions about sarcomere instability on the descending limb of the force–length relationship should be regarded critically and with caution. Sarcomeres are likely to be stable on all portions of their working range, including the descending limb.

In our theoretical considerations on skeletal muscle properties, the instability or softening behaviour of skeletal muscle on the descending limb of the force–length relationship becomes a major problem. We will show mathematically convenient ways of avoiding this problem.

When modelling the human musculo-skeletal system, or parts thereof, it is useful to know not only the generic force–length properties of skeletal muscle, but also importantly, the force–length property of a specific muscle within the anatomical constraints of the skeleton. For example, the human elbow joint can go through an angular displacement of approximately 150°. Questions that are relevant to the force–length properties of elbow flexor muscles include the following. Can the elbow flexors produce force over the entire range of elbow movements? At what joint angles do the elbow flexors produce maximal force?

Modelling skeletal muscle using simple geometric shapes 31

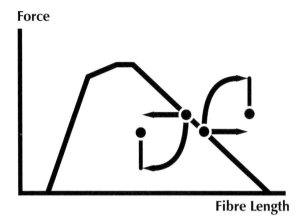

Figure 2.6 Schematic representation of the sarcomere force–length relation of striated muscle. Also shown is the conceptual behaviour of two sarcomeres which are stretched and shortened, respectively, on the descending limb of the force–length relation. Note, that the sarcomeres do not follow the line representing the static force–length relation, and further that the forces in the shortened and lengthened sarcomeres are lower and higher, respectively, than the expected isometric force. For this scenario, sarcomere stiffness is always positive and sarcomere stability is guaranteed.

On what part of the force–length relation do the elbow flexors operate? Are the force–length properties of all elbow flexors about the same, or do they vary significantly? Are the force–length properties of the elbow flexors the same across individuals, or do they adapt to the specific needs of an individual? How are the force–length properties of the elbow flexors influenced by changes in the angles of the neighbouring joints?

It is not our intent to elaborate on all of these questions here; however, some selected comments will be made regarding the operating range of muscles within the constraints of the skeleton, the force–length properties of agonistic muscles (muscles producing moments in the same direction at a given joint), and the variations of force–length properties across subjects.

From the point of view of force production, the optimal operating range of a muscle is centred around the plateau region; i.e. a muscle is working on the plateau region in the middle of the working range, on the descending limb when extremely elongated, and on the ascending limb when extremely shortened (Figure 2.5). Such a working range has been found for a variety of muscles, for example, the frog semimembranosus [63] and the human rectus femoris [46]. However, other muscles have been found to work predominantly on the ascending limb and plateau region, or on the plateau region and descending limb, of the force–length relation (Figure 2.5). For example, during frog jumping the knee and hip angles go from full flexion to full extension, and for this movement the semitendinosus was found to operate primarily on the plateau and descending limb region [65]. Muscles which have been found to work primarily on the ascending limb and the plateau regions include the triceps surae muscles of the striped skunk [33], the triceps surae and plantaris of the cat [43], and the human gastrocnemius [45]. It is not clear why the operating range within the anatomical range of joint movements varies across muscles; however, some insight may be gained by studying changes in force–length properties through chronic alterations of the movement demands (see below).

When modelling joint function, the question arises as to whether or not the force–length properties of all agonistic muscles are similar, and therefore may be represented as a scaled version of the force– (or moment–)angle relation observed for the entire agonistic group. Theoretically, both possibilities have certain advantages. If the force–length properties of all agonistic muscles are similar, then they reach their maximal force potential at a similar joint angle, and therefore the peak forces which can be achieved are high; however, the operating range is restricted. If the force–length properties are not similar and the muscles reach their maximal force potential at different joint angles, the peak forces are not as high; however, the working range is increased compared to the previous possibility (Figure 2.7). Experimental work in this area is sparse; however, it appears that agonistic muscles tend to have similar force–length properties [33, 43]. 'Similar' in the present context refers to the shape of the force–length curve and not to the absolute force values. This result suggests that muscles with similar functions have similar force–length properties, or it might even suggest that the function of a muscle dictates its force–length properties.

In theoretical models of the musculo-skeletal system, it has been assumed (implicitly) that the shape of the force–length property of specific muscles does not vary across subjects [14, 73, 74, 93, 94]. For human or animal subjects who use muscles primarily for normal everyday tasks such as locomotion, force–length properties across subjects are probably similar. However, what about these properties if subjects place different demands on muscles? In a recent study on the force–length properties of the rectus femoris (RF) in elite cyclists and runners it was found that the RF of the cyclists was operating on the descending, and the RF of the runners was operating on the ascending limb of the force–length relation (Figure 2.8). Despite the limited number of observations and the noise in the raw data associated with maximal voluntary contractions in humans, the slope of the best-fitting regression line was negative for all cyclists ($n = 3$) and positive for all runners ($n = 4$) tested, indicating that the force–length properties had adapted to the chronic demands of cycling and running. The adaptations were as one would expect; the cyclists who use the RF in a chronically shortened position, because of the small hip angle associated with cycling, were relatively strong at short compared to long RF lengths; the runners who use the RF at longer lengths than the cyclists, were relatively strong at long compared to short RF lengths. The possible mechanisms of adaptation of the force–length properties in this situation are not known; however, a shift of the force–length relationship in the way observed for these two groups of

Modelling skeletal muscle using simple geometric shapes 33

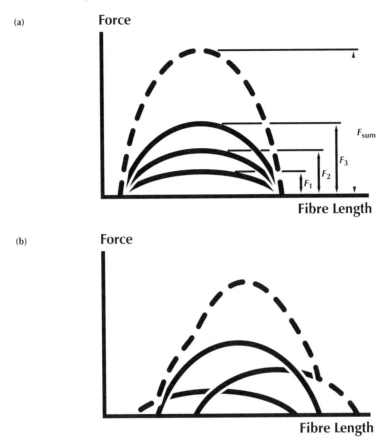

Figure 2.7 Schematic illustration of the total force–length relation (dashed line) of a synergistic group of muscles with (a) similar force–length properties of the individual muscles (solid lines), and (b) with different force–length properties of the individual muscles.

athletes could be achieved through a change in the number of sarcomeres arranged in series in the RF fibres [41]. This possibility is presently being investigated using an animal model of chronic training at short and long muscle lengths. It is hypothesized that training at the short lengths will produce a decrease, and training at the long lengths an increase in the number of in-series sarcomeres in a fibre.

The determination of force–length properties of skeletal muscles in an animal preparation is straightforward. Typically, the distal tendon of the target muscle is removed with a remnant piece of bone, the bone piece is attached to a force transducer, and the muscle is stimulated maximally through its nerve. The

Theoretical models of skeletal muscle

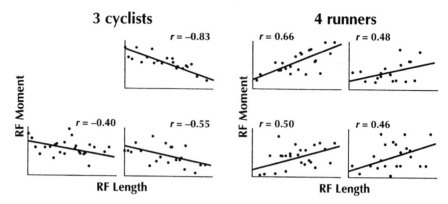

Figure 2.8 Moment–length relations for rectus femoris (RF) of four elite runners and three elite cyclists. Raw data points were approximated with a best-fitting regression line. All regression lines were statistically different from zero, indicating that the RF of the runners operated on the ascending, and the RF of the cyclists operated on the descending limb of the force–length relation. (Adapted and reprinted with permission from [41].)

muscle is then held isometrically at different lengths, and the peak force is plotted as a function of the muscle length [43, 78].

The determination of force–length properties of individual, *in vivo* human skeletal muscles is difficult [45, 46]. Compared to the animal preparation, two basic problems must be considered: how the muscle is stimulated; and how the muscle is isolated from its agonists. Stimulation of *in vivo* human skeletal muscles can be achieved sometimes by using percutaneous electrodes over the target nerve, or by direct percutaneous stimulation of the muscle [10]. These ways of stimulation work well for small muscles, e.g. the first dorsal interosseus [22, 68, 80]; they do not work well for large muscles, because the direct nerve stimulation causes great pain and discomfort and it cannot be used to stimulate the target muscle fully without producing contractions in the neighbouring muscles.

The easiest way of producing muscular contraction is by voluntary effort. Maximal voluntary contractions have been used to determine force–length or moment–angle properties of isolated muscles or groups of muscles. Maximal voluntary contractions are not easy to reproduce, and some muscles cannot be recruited maximally through voluntary efforts [10]. The extent of motor unit recruitment during voluntary contractions can be assessed using the so-called **twitch interpolation** technique (5, 10, 23, 38, 67, 82). In this technique, a single pulse (twitch) stimulation is given to the nerve of the target muscle. The amount of force elicited by the twitch stimulation beyond the voluntarily produced force is taken as an indicator of the extent of muscle activation. Unfortunately, the twitch interpolation technique has a large variance for repeat measurements, and

Modelling skeletal muscle using simple geometric shapes 35

therefore is a qualitative rather than a precise quantitative measure of muscle activation [88].

Isolation of the target muscle from its agonists is another problem which one is faced with when attempting to determine the force–length properties of *in vivo* human skeletal muscles. Isolation procedures have been described for multi-joint muscles which have only one-joint agonists [46]. For example, the human RF muscle is a two-joint muscle crossing the hip (flexor) and the knee (extensor). Its agonists in knee extension are the three vasti, all of them one-joint muscles (Figure 2.9). When performing maximal effort, isometric knee extensor contractions on a strength testing machine, the knee extensor moment can be calculated. This moment is produced by the forces of the three vasti and the RF. In a subsequent contraction, the hip angle is changed and the knee angle is kept the same, thereby altering the length of the RF but not the vasti muscles. A measurement of the knee extensor moment will differ from that of the previous contraction, and the difference can be associated with the change in RF length. When this procedure is repeated for a series of hip angles and a constant knee angle, the shape of the force–length relation of the RF can be determined. There are several non-trivial assumptions underlying this procedure, for example, that the contributions of the knee extensors (and the antagonistic knee flexors) remain the same for changing hip angles. Therefore, the changes in the resultant knee extensor moment are all associated with the RF length change [46].

As an alternative to the methods of maximal nerve stimulation, or maximal voluntary contractions, the force–length properties of an *in vivo* human skeletal muscle could be derived using submaximal nerve stimulation. This procedure has several advantages over the voluntary contractions: for example, it ensures a constant activation of the target muscle. Furthermore, submaximal stimulation reduces the pain and discomfort for the subject. However, submaximally derived force–length properties do not scale linearly with the corresponding maximal

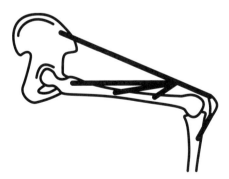

Figure 2.9 Schematic illustration of the human quadriceps muscle group. The three vasti originate from the anterior portion of the femur, the rectus femoris originates from the anterior superior iliac spine of the pelvis.

Theoretical models of skeletal muscle

properties [37, 39, 78]. In summary, it is not trivial to determine the force–length properties of *in vivo* human skeletal muscles.

So far, only the active force–length properties of skeletal muscle have been discussed. When a muscle is pulled to increasingly longer lengths, passive resistance is encountered which is associated with elastic elements arranged in parallel with the contractile elements of the muscle. In some muscles, this resistance is considerable even at fibre lengths corresponding to the optimal sarcomere length (e.g. in cardiac muscle). In the limb musculature of humans and animals, passive muscular forces usually play a minor role within the normal anatomical range of joint movement. The properties of the passive elements of skeletal muscle are discussed in Section 2.5.

2.3.2 Force–velocity properties of skeletal muscle

The second property of skeletal muscle which is typically associated with the contractile element of SLMs is the force–velocity relation. The force–velocity relation describes the maximal, steady-state force of a muscle as a function of its rate of change in length. A maximally stimulated muscle has a decreasing force potential with increasing velocities of shortening, and an increasing force potential with increasing velocities of elongation (Figure 2.10). Hill (48) described the force–velocity relation for shortening muscle at optimal length using part of a rectangular hyperbola:

$$(F + a)(v + b) = (F_0 + a)b \qquad (2.7)$$

where F is the steady-state force for shortening at a velocity v; F_0 is the maximal,

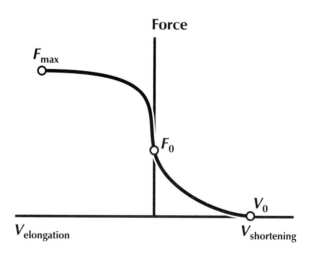

Figure 2.10 Schematic illustration of the force–velocity relation of skeletal muscle.

Modelling skeletal muscle using simple geometric shapes

isometric force at optimal contractile element length, and a and b are constants with units of force and velocity, respectively.

Solving for F, equation (2.7) becomes

$$F = \frac{F_0 b - av}{b + v} \quad (2.8)$$

F can be calculated if F_0, v, a, and b are known. For shortening at the maximal velocity, v_0, the force F becomes zero (Figure 2.10), therefore equation (2.8) may be rewritten for this special case as

$$F_0 b = av_0$$

or

$$a/F_0 = b/v_0 = 0.25 \quad (2.9)$$

a/F_0 and b/v_0 are dimensionless quantities approximately equal to 0.25 for many muscles across species and temperatures [50], including human fast-twitch fibres at 37°C [28, 29].

Force–velocity properties are available for a variety of animal muscles (20, 24, 26, 48, 58, 85), and moment–angular velocity relations have also been derived for agonistic groups of human skeletal muscles during maximal voluntary contractions [76, 92]. Force–velocity descriptions of individual human skeletal muscles are rare. Faulkner et al. [28, 29] determined force–velocity properties of isolated fibre bundle segments (10–25 mm long) from human fast- and slow-twitch muscles and found values for a/F_0 of 0.25 and 0.15 for fast and slow fibres, respectively. Therefore, knowing the physiological cross-sectional area of a muscle, the maximal isometric force, F_0 may be calculated using equation (2.3), and the constant a may then be calculated as $a = 0.25F_0$.

For a given cross-sectional area, slow- and fast-twitch fibres produce similar levels of isometric force, but the maximal velocity of shortening differs by a factor of 3 or 4. For human skeletal muscles, Faulkner et al. [28] reported values for v_0 of 6 and 2 fibre lengths per second for fast and slow fibre bundle preparations, respectively. Compared to other reports of mammalian skeletal muscle at 37°C, these v_0-values appear low. Values obtained from fast and slow muscles in the mouse and rat range from 9.4 to 24.2 lengths per second and from 6.5 to 12.7 lengths per second, respectively [18, 19, 21, 62, 79]. One of the reasons why the human v_0-values reported by Faulkner et al. [28] are relatively low may be associated with the fact that they did not measure v_0 but approximated v_0 by extrapolating Hill's equation to a value of zero force. Edman [24] showed that Hill's equation gives good approximations of experimental force–velocity relations in a range of approximately 5–80% of the isometric force, but that it overestimates the actual values of F_0 and underestimates the actual values of v_0 considerably.

Measuring the maximal velocity of shortening in human skeletal muscles is difficult; however, a rough estimate may be obtained by determining the maximal power output of a group of muscles as a function of movement speed, and knowing that maximal power is achieved at 31% of v_0, if it is assumed that $a/F_0 = b/v_0 = 0.25$ [40]. For human knee extensor muscles, Suter et al. [89] found that maximal power was produced at knee angular velocities of about 240°/s and 400°/s for predominantly slow-twitch fibred subjects (more than 55% slow-twitch fibres in the vastus lateralis) and predominantly fast-twitch fibred subjects (more than 60% fast-twitch fibres in the vastus lateralis). Therefore, the maximal speeds of contractions for these two subject groups are approximately 774°/s and 1290°/s, respectively. Assuming a moment arm of the knee extensor muscles about the knee axis of 0.05 m [44] at the instant of peak force production, the speeds of shortening for the two groups are about 0.68 m/s and 1.13 m/s. Knowing that the average fibre length in the knee extensor muscles is approximately 0.08 m [101], the maximal speeds of shortening expressed in terms of the contractile element lengths are about 8 and 14 lengths per second for the slow-twitch and the fast-twitch fibred groups, respectively. These values for v_0 agree better with those obtained for slow and fast mammalian skeletal muscles [18, 19, 21, 62, 79], and they are considerably larger than those obtained by Faulkner et al. [28] for human slow and fast fibres. The estimated v_0-values of 8 and 14 lengths per second must be considered with some reservation because they were extrapolated from measurements of mixed muscles. It is expected that the v_0-value of a purely fast-twitch fibred muscle should be higher than 14 lengths per second, and the v_0-value of a purely slow-twitch fibred muscle should be lower than 8 lengths per second.

Once an estimate of F_0 and v_0 is obtained, the constants in Hill's [48] equation, a and b, can be calculated using equation (2.9). Knowing F_0, v_0, a, and b, the maximal, steady-state force, F, as a function of the shortening velocity, v, at optimal contractile element length is given by equation (2.8).

So far, we have primarily discussed the force–velocity properties of shortening muscle. When a muscle is stretched at a given speed, its force exceeds the maximal isometric force, F_0, reaching an asymptotic value of about $2F_0$ at speeds of stretching much lower than the maximal velocity of shortening [61]. Also, using isotonic [58] or isokinetic stretches [25], there appears to be a discontinuity in the force–velocity relation across the isometric point: the rise in force associated with slow stretching is much larger than the fall in force associated with the corresponding velocities of shortening. In contrast to the force–velocity relation during shortening, the force–velocity relation during stretching is rarely described using a standard equation, such as the hyperbolic relation proposed by Hill [48] for shortening. The primary reason for this discrepancy is the fact that force–velocity properties during stretching have been investigated much less, and that these properties are not as consistent as those obtained during shortening. For example, the force–velocity relations appear to depend on the type of experiment: isotonic or isokinetic.

In isotonic experiments [58], forces greater than F_0 are suddenly applied to the fully activated muscle and the corresponding speeds of stretching are recorded.

Modelling skeletal muscle using simple geometric shapes

These speeds follow a complex time course: first, there is a quick (instantaneous) lengthening which is typically associated with the series elasticity in the muscle, then there is a fast and finally a slow lengthening of the fibres. The final, slow fibre lengthening is thought to reflect the actual contractile behaviour of skeletal muscle, and it is this lengthening speed that produces the discontinuous behaviour of the force–velocity relation across the isometric point.

In isokinetic experiments, the velocity of stretch is controlled and the forces are measured. Depending on the speed of stretch, the corresponding forces continue to increase throughout the stretch (slow speeds), the forces reach a plateau (intermediate speeds), or the forces reach a peak and fall for the remainder of the stretch (fast speeds) [25]. Depending on the force which is taken as the representative force for the stretch experiment, different force–velocity relations may be obtained. The experimental difficulties associated with determining a consistent force–velocity behaviour of skeletal muscles during stretching have resulted in the current state of affairs in which the behaviour of muscle during stretch is much less known (and described) than the behaviour of muscle during shortening.

When modelling the force–velocity properties of skeletal muscles, the fibre type distribution must be accounted for [28, 50, 64]. Slow- and fast-twitch fibres of the same cross-sectional area have similar F_0-values, but they differ substantially in the maximal velocity of shortening (Figure 2.11a). Skeletal muscles are typically comprised of a mixture of slow- and fast-twitch fibres. Hill [50] made an attempt to model the force–velocity properties of a muscle containing 82 'fibres' with 10 different intrinsic speeds of shortening. He found that the regular force–velocity equation (equations (2.7), (2.8)) fitted the mixed-fibred muscle well, except at forces less than 5% of F_0. Here, the speeds of shortening of the mixed-fibred muscle exceeded the corresponding predicted values considerably. It was argued that the maximal speed of shortening in a mixed-fibred muscle could not be predicted well using Hill's [48] equation, because the high-speed portion of the experimental relation was dominated by fibres which had a much higher than average speed of shortening.

The difference in fibre type distribution of skeletal muscles is particularly important for power production. Power (P) is defined as

$$P = \mathbf{F} \bullet \mathbf{v} \qquad (2.10)$$

where $\mathbf{F}$ is the force (vector), $\mathbf{v}$ is the velocity (vector) and • represents the scalar (or dot) product. For muscles, power is calculated as the product of the muscle's force magnitude (scalar) and the speed of contraction (scalar). The power of a fast-twitch fibre is higher at all speeds of shortening than that of the corresponding slow-twitch fibre (Figure 2.11b). Also, the peak power is achieved at a higher absolute speed of shortening in the fast than in the slow fibre.

In a study aimed at determining the power output of slow and fast human skeletal muscle fibres, Faulkner *et al.* [28] found that slow fibres of isometric force equal to that of fast fibres produced peak powers that were 25% of those of

40 Theoretical models of skeletal muscle

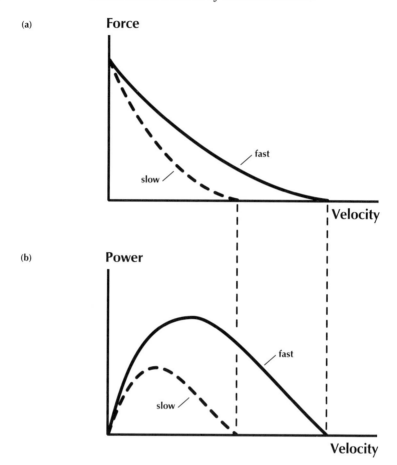

Figure 2.11 Schematic illustration of (a) the force–velocity properties, and (b) the power–velocity properties of a fast and a slow skeletal muscle fibre.

the fast fibres (Figure 2.12). Also, a muscle comprised of 50% slow- and 50% fast-twitch fibres was estimated to have a peak power value of only 55% of that of a corresponding 100% fast-twitch fibred muscle. Similar results were found by other investigators [64].

Up to now, the force–length and the force–velocity properties have been treated as separate entities. Once both these properties have been determined, the question arises as to how they should be combined in a description of contractile element properties. It has been suggested that Hill's [48] equation of the force–velocity relation (equations (2.7) and (2.8)) can be used for all muscle lengths provided that the F_0-value (the maximal isometric force at optimal length) is replaced by the maximal isometric force at the length of interest

Modelling skeletal muscle using simple geometric shapes 41

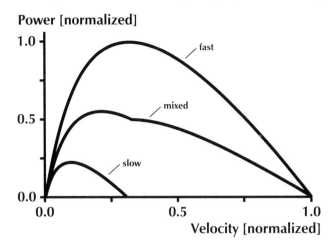

Figure 2.12 Power–velocity relation of human fast, slow, and mixed (50% fast, 50% slow) fibred skeletal muscle. (Adapted from [28].)

($F_0(l)$, [2]). Using this suggestion, equation (2.8) becomes

$$F = \frac{F_0(l)b - av}{b + v} \quad (2.11)$$

Assuming that v is zero (i.e. we have an isometric contraction) $F = F_0(l)$, which is correct. The maximal velocity of shortening, v_0, according to equation (2.11), becomes

$$v_0 = \frac{F_0(l)b}{a} \quad (2.12)$$

because F is zero in this situation. Since a and b are constants in equation (2.12), v_0 depends directly on $F_0(l)$. This result suggests that the maximal speed of shortening is highest at optimal muscle length and becomes progressively smaller at lengths other than optimal. However, Edman [24] showed convincingly that v_0 does not depend directly on the contractile element length. He performed experiments on single fibres of frogs aimed at measuring v_0 as a function of sarcomere length. The results of these experiments are summarized in Figure 2.13. v_0 was nearly constant in the range of sarcomere lengths from 1.65 μm to 2.70 μm. Below sarcomere lengths of 1.65 μm, v_0 fell because of a presumed force resisting shortening at these sarcomere lengths [32]. Beyond sarcomere lengths of 2.70 μm, v_0 increased with increasing sarcomere lengths. This increase was associated with passive elastic forces which produce rapid shortening of unstimulated muscle fibres at these lengths and which appear to enhance v_0 in the stimulated fibre.

Since v_0 appears to be constant for a large range of contractile element length,

42 Theoretical models of skeletal muscle

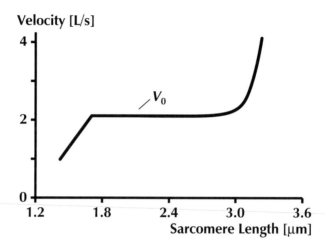

Figure 2.13 Maximal velocity (v_0) of shortening of frog skeletal muscle fibres as a function of sarcomere length. Note that v_0 remains nearly constant for sarcomere lengths ranging from 1.65 to 2.70 μm. (Adapted from [24].)

a better representation of the force–length–velocity relation than that shown in equation (2.11) may be achieved by multiplying Hill's [48] force–velocity relation by a normalized length factor, at least for contractile element lengths ranging from 1.65 to 2.70 μm:

$$F = \left(\frac{F_0 b - av}{b + v}\right) f(l) \qquad (2.13)$$

where $f(l)$ represents the normalized force according to the force–length properties of the muscle; $f(l)$ ranges from 0 to 1.0. For sarcomere lengths outside 1.65–2.70 μm, a different representation of the force–length–velocity relation is required.

Combining the force–velocity and force–length relationships also causes a conceptual problem. By definition, the force–length properties of skeletal muscles are determined under isometric conditions by fully activating the muscle at a series of discrete lengths. Therefore, the force–length relationship consists of a series of discrete force measurements obtained at discrete muscle lengths. In contrast to the force–length property, the force–velocity relationship is obtained for a series of dynamic contractions, typically performed under isotonic or isokinetic conditions. When combining the two relationships, for example, as shown in equations (2.11) and (2.13), the force–length property is treated like a continuous function. This treatment of the force–length property causes (at least) two conceptual problems. First, force–length relationships are valid for isometric contractions; equations (2.11) and (2.13), however are aimed at describing

Modelling skeletal muscle using simple geometric shapes

dynamic processes. Secondly, the force–length property is treated like an instantaneous rather than a steady-state property which causes stability problems on the descending part of the force–length relationship, as discussed previously. These issues will be considered further in the theoretical parts of this book.

2.3.3 History-dependent properties of skeletal muscle

History-dependent properties could mean a variety of phenomena. Here, we are concerned with the force depressions following shortening and force enhancement following stretch of a skeletal muscle.

Force depression following shortening

It is well accepted that the isometric force following shortening of a muscle, or a single fibre, is smaller than the corresponding force of an entirely isometric contraction [1, 66, 87]. That is, if a muscle is fully activated at a given length, l_1, and then is allowed to shorten to a new length, l_2, at which the fully activated isometric force is measured, then the force at l_2 following shortening will be smaller than the force obtained for a purely isometric contraction at l_2. This force depression following shortening is directly related to the amount of shortening (Figure 2.14a), and is inversely related to the speed of shortening (Figure 2.14b) [1, 66, 87].

During and after shortening contractions, the decrease in force has been paralleled by a corresponding decrease in fibre stiffness [87]. If it is assumed that stiffness is an indicator of the number of force-producing sites (cross-bridges) in the muscle [30, 52, 53, 57], then the decrease in stiffness during and after shortening indicates a decrease in the number of force-producing sites. At present, it is not clear why the number of force-producing sites should be decreased for long periods (several seconds) following shortening. It has been proposed that there might exist a stress-related inhibition of the force-producing sites which enter the overlap region between the thick and thin myofilaments during shortening contractions [66]. A stress-related inhibition is an attractive theory, because it could account for the increased force depression with increasing shortening distance (i.e. more force-producing sites come into the overlap zone for a large than for a small shortening distance). It could also explain the increased force depression with decreasing shortening speeds (slow speeds of shortening are associated with a larger stress than fast speeds of shortening).

Another mechanism which has been proposed to explain the force depressions following shortening is sarcomere (or segment) length non-uniformities [49]. Sugi and Tsuchiya [87] found that sarcomere length changes in fibres of frog tibialis anterior were reasonably uniform at fast speeds of shortening, but were highly non-uniform at slow speeds of shortening, therefore supporting the argument that the force depressions following shortening may be associated with

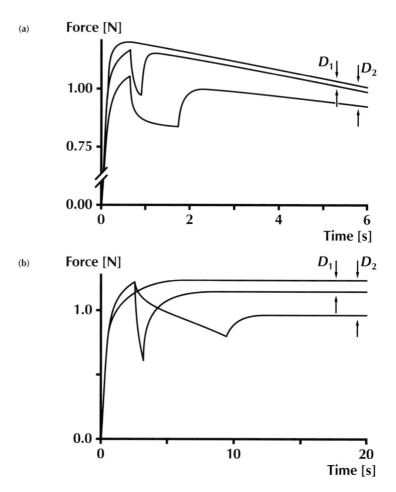

Figure 2.14 (a) Force depression following shortening increases with increasing shortening distances (i.e. $D_2 > D_1$). Top trace: isometric contraction at optimal length for frog skeletal muscle. Middle trace: shortening of 0.56 mm to the optimal length at 2.5 mm/s. Bottom trace: shortening of 2.8 mm to the optimal length at 2.5 mm/s. (Adapted and reprinted with permission from [66].) (b) Force depression following shortening increases with decreasing speeds of shortening (i.e. $D_2 > D_1$). Top trace: isometric contraction at a muscle length of 37 mm in dogfish jaw muscle. Middle trace: shortening from 42 to 37 mm at a speed of 8 mm/s. Bottom trace: shortening from 42 to 37 mm at a speed of 0.5 mm/s. (Adapted from [1].)

Modelling skeletal muscle using simple geometric shapes 45

increased sarcomere non-uniformities. However, a variety of observations appear to weaken this argument. For example, the fact that the force depressions depend on the amount of shortening, which, for a given speed is associated presumably with a similar non-uniformity in sarcomere length [87], does not fit the non-uniformity argument. Furthermore, the non-uniformities in sarcomere lengths persist for a long time (seconds) following the shortening step, suggesting that the sarcomeres produce similar forces despite their non-uniform lengths [34, 87]. Since the force following shortening is depressed compared to the corresponding purely isometric force, it is safe to assume that the sarcomeres at lengths with a great force potential are inhibited (in some way) compared to the sarcomeres at lengths with a small force potential. This argument fits nicely with the idea of a stress-induced inhibition of force production following shortening contractions. Also, in segment clamped experiments (i.e. experiments in which the sarcomere lengths of one region of the fibre are precisely controlled) in which sarcomeres shortened uniformly and remained at uniform lengths following shortening, the same force depression as that observed in non-uniform preparations was observed, indicating that non-uniform sarcomere lengths cannot be the primary reason for the force depressions following a shortening step [34].

Force enhancement following stretch

As for the force depressions following shortening, it is well accepted that the long-lasting isometric force following stretch of a muscle or fibre is larger than the corresponding force of an entirely isometric contraction [1, 25, 87]. Force enhancement following stretch increases with increasing amplitudes of stretch [25], appears to be independent of stretch velocities for single fibres [25, 87], but appears to depend on stretch velocities for whole muscles [1] (Figure 2.15).

Force enhancement following stretch is most pronounced on the descending limb of the force–length relation [25], (i.e. at sarcomere lengths longer than optimal). Force enhancement increases for a given amplitude of stretch with sarcomere length, thus the force following stretch is not proportional to the overlap of thick and thin myofilaments [25]. Finally, force enhancement following stretch is associated with a muscle/fibre stiffness of similar magnitude to the stiffness measured during the corresponding isometric contractions. This last result suggests that the force enhancement is accomplished through an increase in the average force of a force-producing site (cross-bridge), rather than through an increase in the number of force-producing sites. Also, sarcomere lengths remain relatively uniform during and following stretches [87], and so may not account for the force enhancement.

The mechanisms underlying force enhancement following stretch are not known. From the experimental observations, some speculative hypotheses may be proposed. Amemiya *et al.* [7] found that slow stretching of stimulated frog skeletal muscle produced a disorder in the myofilament lattice which could result

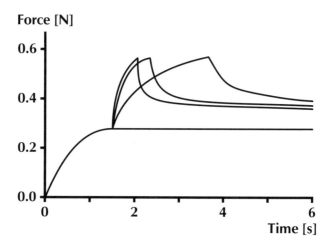

Figure 2.15 Force enhancement following stretch increases with decreasing speeds of stretch. Bottom trace: isometric contraction of toad sartorius (0°C) at a length of 25.5 mm. Top traces: stretching of the same muscle from 21.5 to 25.5 mm at 1.9 mm/s (top), 5 mm/s (middle), and 8 mm/s (bottom trace). (Adapted from [1].)

in an increase in the isometric force following stretch. Although this finding may provide a qualitative explanation for force enhancement, it appears doubtful that it could account for the large force enhancements observed experimentally.

Since muscle stiffness of entirely isometric contractions is about the same as the stiffness following stretch to the corresponding length, it follows that the force enhancement is associated with an increased force per force-producing site. According to the cross-bridge theory, such an increase in force seems only possible if the force-producing sites are stretched further from the equilibrium position after elongation of the muscle compared to the isometric condition [52, 53]. This explanation of the force enhancement appears feasible for short time periods following the stretch; however, the cyclic attachment–detachment of the force-producing sites [52] should produce forces after stretch equal to those of isometric contractions, except if it was postulated that the force-producing sites have different attachment characteristics following stretch than they have during entirely isometric contractions. Such a speculation has not been tested to date.

Stretch–shortening versus shortening–stretch

In order to test whether the history-dependent properties of skeletal muscles following shortening and stretch are commutative, we performed stretch–shortening and shortening–stretch experiments using cat soleus. Figure 2.16 shows the results for three stretch–hold–shortening experiments (a, a′, a″) and the

Modelling skeletal muscle using simple geometric shapes 47

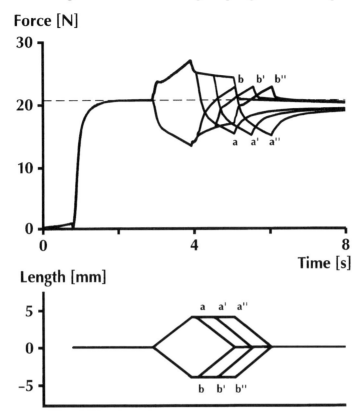

Figure 2.16 Stretch–hold–shortening (a, a′, a″) and shortening-hold-stretch (b, b′, b″) tests in cat soleus at 34.0°C. Stretching and shortening were performed for a distance of 4 mm at a speed of 4 mm/s. All contractions started and finished at the optimal length of the muscle. The hold times were 50 ms (a, b), 550 ms (a′, b′) and 1050 ms (a″, b″). Note, that the initial isometric forces are virtually identical in all tests, while the final isometric forces at optimal length were always higher for the shortening–stretch than the stretch–shortening tests.

corresponding shortening–hold–stretch experiments (b, b′, b″). Note, that independent of the order of stretching and shortening, the initial and final isometric muscle length was the same in all six tests. However, the isometric force is only the same before the length changes were performed. Following the length changes, the isometric forces in the stretch–shortening experiments (a, a′, a″) were always lower than the corresponding forces in the shortening–stretch experiments (b, b′, b″). The same result was obtained when the stimulation was reduced to submaximal levels during the hold phase by decreasing the frequency of nerve stimulation (Figure 2.17).

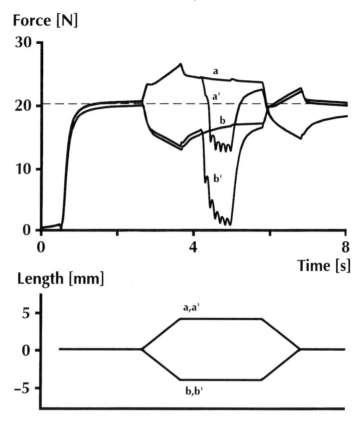

Figure 2.17 Stretch–hold–shortening (a, a′) and shortening–hold–stretch (b, b′) tests in cat soleus at 34.0°C. Stretching and shortening were performed for a distance of 4 mm at a speed of 4 mm/s. All contractions started and finished at the optimal length of the muscle. The hold times were about 2100 ms. In tests a and b, soleus nerve stimulation was maximal, while in tests a′ and b′, nerve stimulation was reduced to a frequency of 7 Hz for an 800 ms period. Note that the initial isometric forces are similar while the final isometric forces at optimal length were always higher for the shortening–stretch than the stretch–shortening tests.

2.4 ANGLE OF PINNATION

In the previous sections of this chapter, the force-producing properties of the contractile element were discussed. One of the implicit assumptions of most models of skeletal muscle, and all SLMs, has been that the line of action of the contractile element force follows the long axis of the fibres. In SLMs, the line of action of the contractile element is straight, therefore knowledge of the fibre direction is imperative.

The fibre direction relative to the line of action of the muscle [103], or relative to the aponeurosis (tendon sheath – e.g. [72]), can be described by a single

Modelling skeletal muscle using simple geometric shapes 49

parameter, the angle of pinnation (Figure 2.18). In multipennate muscles, the number of angles required to describe the geometry of the muscle equals the number of distinctly different fibre directions.

Most often the angle of pinnation has been determined in cadaveric specimens, and changes in the angle of pinnation as a function of muscle length or force were ignored. For example, Wickiewicz *et al.* [101] measured the angles of pinnation of 27 human lower limb muscles in three cadaveric specimens. The angle of pinnation was defined by the line of action of the muscle and the fibre direction in the anatomic position. Only one angle of pinnation estimate was given for each muscle, even for muscles classified as bi- or multipennate. Wickiewicz *et al.* [101] also discussed the physiological implications of the muscles' architectural design (force production, range of force production, maximal velocity of shortening), disregarding the possible influence a changing angle of pinnation might have on these properties.

Wagemans and van Leemputte [96] were probably the first to show that angles of pinnation changed systematically as a function of muscle length and force production in *in vivo* human gastrocnemius muscles. In that study, 21 subjects performed isometric ankle plantar flexor contractions at five different ankle angles. Ultrasound images of the medial head of the gastrocnemius taken during the contraction revealed that angles of pinnation increased with increasing force production at a given muscle length, and also increased with shortening of the muscle at a given (zero) force.

The results of Wagemans and van Leemputte [96] were supported in part by studies in the rat medial gastrocnemius muscle. Zuurbier and Huijing [109] found an almost linear increase in the angle of pinnation of the most distal fibre with decreasing muscle length during isometric contractions. In a similar study by the same group [111], the angle of pinnation of the most distal fibre of the rat gastrocnemius was found to vary from about 15° (at optimal length + 2 mm) to about 45° (at optimal length −8 mm), thus indicating that changes in the angle of pinnation may contribute substantially to the length change of the muscle (Figure 2.19).

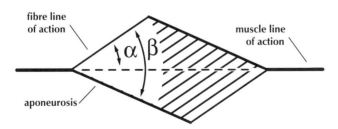

Figure 2.18 The angle of pinnation defined as the angle between the line of action of the fibre and the line of action of the muscle, α; and as the angle between the line of action of the fibre and the aponeurosis, β.

50   *Theoretical models of skeletal muscle*

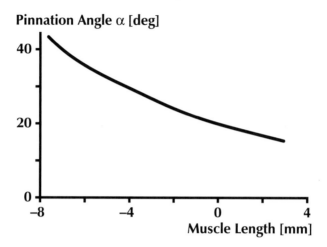

Figure 2.19 Changes in the angle of pinnation as a function of muscle length for the most distal fibre in the rat gastrocnemius muscle. 0 mm indicates the optimal length of the muscle. (Adapted from [111].)

Brooks *et al.* [15] found similar results to those reported above in a study on the unipennate cat medial gastrocnemius muscle. In addition, these researchers found up to 10° differences in the angles of pinnation, and up to 33% length changes of fibres when the muscle went from a relaxed to a fully active state during isometric (i.e. fixed muscle–tendon length) contractions (Fig. 2.20). Furthermore, they found a direct proportional relation between the angle of pinnation and the muscle force during isometric contractions. Evaluation of the angles of pinnation from five fibres along the cat medial gastrocnemius also revealed systematic differences (up to 20°) in the angles of pinnation as a function of location within the muscle.

In most models of skeletal muscle, the angle of pinnation as a function of force and muscle length is not known. However, using the assumption that muscle thickness does not vary with force and muscle length [4], the relation between thickness (t), fibre length (L_f) and the angle of pinnation (β) for a unipennate muscle becomes (Figure 2.21):

$$L_f = \frac{t}{\sin(\beta)} \quad (2.14)$$

Benninghof and Rollhäuser [12] calculated the relation between fibre length (L_f), angle of pinnation (α), and shortening distance (s) for a pennate muscle with non-uniform fibre lengths and angles of pinnation, assuming that muscle volume remains constant. They accounted for non-uniform fibre lengths and angles of pinnation by requiring that the point of insertion of all fibres moves the same

Modelling skeletal muscle using simple geometric shapes 51

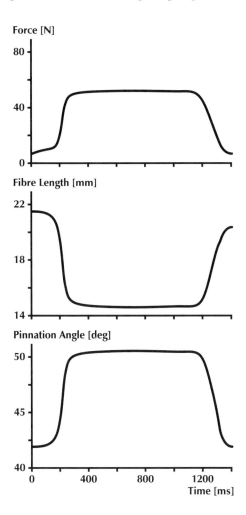

Figure 2.20 Force–time, fibre length–time, and angle of pinnation–time histories of an isometric contraction (i.e. the end-points of the muscle–tendon complex were fixed) of cat medial gastrocnemius muscle at an ankle flexion angle of 139° (i.e. a length corresponding to the shortest muscle length during normal locomotion).

distance when the fibres shorten the same relative amount:

$$L_\mathrm{f} = \frac{s}{\cos(\alpha) - \sqrt{\cos^2(\alpha) + (L_\mathrm{f}/L_\mathrm{fo})^2 - 1}} \quad (2.15)$$

where L_fo is the fibre optimal length.

Constancy of muscle volume [12, 31] or constancy of volume of segments of

52 Theoretical models of skeletal muscle

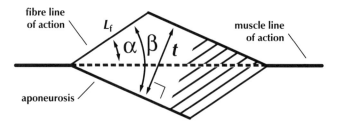

Figure 2.21 Schematic representation of the relation between fibre length, L_f, muscle thickness, t, and the angle of pinnation, β.

muscle [103] has been a primary assumption to define structural changes in muscles during contraction, or to calculate pressures within muscles [47, 72]. The fact that muscle volumes remain nearly constant during contractions was asserted over 300 years ago by experiments on frog and human muscles (Swammerdam, 1663, cited in [69]), and has been confirmed more recently to be accurate to within a fraction of a percent [9]).

Summarizing, it is important to realize that the direction of fibre forces can be given using the angle of pinnation in SLMs. The angle of pinnation typically (but not always) increases with decreasing muscle lengths, and increases with increasing force for a given length [15, 96, 109, 111]. The angles of pinnation found in studies involving fixed specimens rarely exceed 25–30° [83, 101], whereas angles of pinnation of 45° and changes in these angles of up to 30° can be found in *in situ* or *in vivo* experiments [15, 96, 109, 111].

2.5 PASSIVE ELEMENT PROPERTIES

2.5.1. Passive elements in parallel with the contractile element

When a relaxed muscle is stretched, it develops passive force as a function of length. The term 'passive' refers to the fact that this type of force, in contrast to the active force, does not require metabolic energy. The parallel passive forces are associated with connective tissue structures surrounding the muscle fibres, fascicles, and the entire muscle. They are thought to be primarily elastic [102, 103].

In SLMs of muscles, the parallel elastic elements are either ignored, because they are assumed to play a minor role in force production over the range of physiologically relevant muscle lengths, or they are represented as part of the force–length property of the muscle [103]. Parallel elastic element properties have been determined for a variety of muscles; they appear to vary significantly across muscles in terms of force–elongation properties, and in terms of where (relative to the active force–length relation) significant non-zero passive forces can be detected. For example, the contribution of passive forces to muscular

Modelling skeletal muscle using simple geometric shapes

forces occurs on the ascending limb of the force–length relation in the frog gastrocnemius, whereas the onset of the passive force occurs at about optimal length and on the descending limb of the force–length relation for the frog sartorius and the frog semitendinosus, respectively (Figure 2.22). This example demonstrates that the passive elastic element properties must be determined for each muscle individually. Typically, the passive elastic forces are algebraically added to the active forces at the corresponding lengths to give the total (i.e. active plus passive) force-producing ability of the muscle (Figure 2.22).

2.5.2 Passive elements in series with the contractile element

By their morphological arrangement, the in-series passive elements transmit the forces from the muscle (the active and the parallel elastic forces) to the bone. The in-series passive elements are associated with linearly elastic elements in the cross-bridge [52, 53] and the myofilaments [54, 55, 70, 97], as well as with elastic or viscoelastic tendons and aponeuroses [84, 105, 108]. For SLMs, only the properties of the tendons and aponeuroses (Figure 2.23) are of interest; possible cross-bridge or myofilament elasticity is assumed to be accounted for in the active properties of the contractile element.

Tendons

Tendon force–elongation or stress–strain properties differ among muscles [11, 106]; however, some features appear preserved across tendons. Force–elongation curves of tendons are typically nonlinear; they have a low

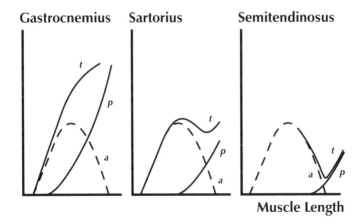

Figure 2.22 Force–length properties of frog gastrocnemius, sartorius, and semitendinosus muscles; t is the total force, p the passive force, and a the active force. (Adapted from [102].)

54   *Theoretical models of skeletal muscle*

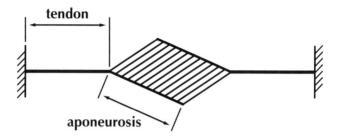

Figure 2.23 Schematic illustration of the tendon and aponeurosis elements in unipennate muscle.

stiffness toe region [81] where stresses rarely exceed 10 MPa for strains of 1–3%, followed by a region (3–6% strain) of substantially increasing stiffness, reaching tangent elastic moduli in the order of 1.0–1.5 GPa (Figure 2.24). Failure strains of tendons are approximately 8–10%.

Another general feature of tendons is their slight viscoelasticity. For cyclic loading–unloading, the energy released by tendons in the unloading phase is about 85–95% of the energy received during the corresponding loading phase (Figure 2.24). Because of the small loss in energy, tendons are well suited to store potential energy during loading of the muscle–tendon complex, and release much of this energy as mechanical work when the muscle–tendon complex is unloaded [3, 17, 59].

Most force–elongation or stress–strain properties of tendons have been determined using a single strain measurement. Recently, it has been found that strains along tendons or ligaments [16, 71, 86, 107] and across tendons [8] are non-

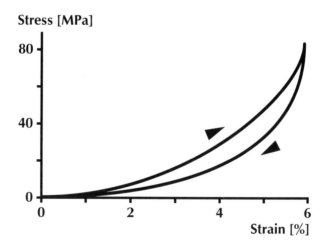

Figure 2.24 Schematic stress–strain curve of tendinous tissues.

uniform. Specifically, surface strains in the middle portion of tendons were found to be smaller than the corresponding strains close to the insertion sites. Although such non-uniformities in strain behaviour may have little consequence on the modelling of the fibre–tendon interactions, local strain measurements may over- or underestimate the actual strain (elongation) of the entire tendon, resulting in errors of the estimates for contractile element length and speed.

The influence of tendon elasticity on force production of an SLM may be illustrated as follows. Imagine a parallel-fibred muscle with a rigid, and an elastic, tendon. For the same contraction of the contractile element, the muscle–tendon length of the muscle with the elastic tendon is larger than that of the muscle with the rigid tendon because of the force-induced stretch on the elastic tendon (Figure 2.25c). Viewing this same situation from identical muscle–tendon lengths, the contractile element would always be shorter in the muscle with the elastic compared to the muscle with the rigid tendon (Figure 2.25b). Plotting the force–length relation of these two muscles (force versus muscle-tendon length, Figure 2.25a), the relation of the muscle with the elastic tendon is shifted to longer muscle lengths (i.e. to the right along the length axis) compared to the corresponding relation of the muscle with the rigid tendon because for a given contractile element length (and thus, muscle force), the muscle–tendon length is always longer for the muscle with the elastic tendon than for the muscle with the rigid tendon (Fig. 2.25a). Clearly, the difference in the tendon elasticity of the two muscles influences the active force–length relation, although the contractile element properties are identical.

Aponeuroses

In SLMs, aponeuroses are typically treated as rigid elements [103]. However, the results of recent studies indicate that aponeuroses have elastic properties [84, 110]. From the few quantitative works in which the elastic properties of tendon and aponeuroses were determined simultaneously in the same muscle, it appears that aponeuroses are as compliant as or more compliant than tendons [27, 84, 108], and that non-uniformities of aponeuroses' strains are larger (between 3–5% strain in the middle portion and over 50% strain in the most distal portion of the rat medial gastrocnemius aponeurosis [108]) than corresponding non-uniformities in tendons (16, 71, 86, 107).

Aponeuroses and tendons have a wave-like 'crimp' pattern in the unstretched state. This crimp pattern is thought to disappear with initial elongation of the series elastic elements without appreciable force production. Therefore, aponeuroses and tendons may elongate at the beginning of muscular contractions without transmitting forces to the bony attachment sites.

Summarizing, the passive properties of skeletal muscles are elastic, or nearly so. The properties differ among muscles, and therefore must be determined for each muscle independently. Parallel elastic forces may influence total muscle force production considerably within the normal anatomical range of lengths.

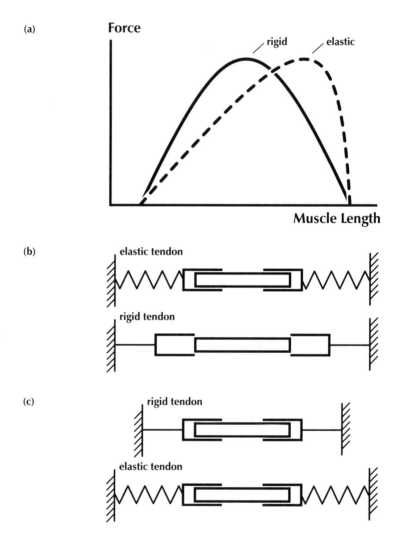

Figure 2.25 (a) Schematic force–length relation of muscles with rigid and elastic tendons. (b) For a given force and muscle–tendon complex length, the contractile element of the muscle with the elastic tendon will always be shorter than the contractile element of the muscle with the rigid tendon, provided the muscle force is greater than zero. (c) For a given force and contractile element length, the length of the muscle–tendon complex for the muscle with a rigid tendon will always be smaller than the length of the muscle with the elastic tendon, provided the force is greater than zero.

Modelling skeletal muscle using simple geometric shapes

The in-series (visco)elastic properties of tendons and aponeuroses influence the variable contractile element length and speed, and therefore its force-producing ability. Parallel elastic and series (visco)elastic properties of passive structures should be considered in models of skeletal muscle.

PROBLEMS

2.1 Imagine two muscles of equal volume. Muscle A has fibres of twice the length of those of muscle B, and the physiological cross-sectional area of muscle A is one-half of that of muscle B. The two muscles have the same distribution of muscle fibre types.

Answer and justify the following questions:

(a) Which of the two muscles has the larger maximal isometric strength?

(b) Which of the two muscles can exert active force over a larger range of lengths?

(c) Which of the two muscles has the larger work potential? Work potential, here, refers to the maximal ability of the muscle to produce work during a single contraction.

Draw the following schematic graphs, and justify your representations of muscles A and B.

(d) Draw a schematic force–length relation of the two muscles.

(e) Draw a schematic force–velocity relation of the two muscles.

(f) Draw a schematic power–velocity relation of the two muscles.

Solution

(a) Muscle B has the larger maximal isometric strength because of the larger physiological cross-sectional area of muscle B than of muscle A.

(b) Muscle A can exert active force over a larger range of lengths than muscle B because of the larger fibre lengths of A compared to B.

(c) The two muscles have the same work potential, because they have equal volumes. Equal volumes implies that both muscles have the same number of sarcomeres, and, assuming each sarcomere has the same work potential, the work potential of the muscles is equal.

(d) Muscle B is stronger than A (as explained in answer a), and muscle A has the larger range of active force production (as explained in answer b); see Figure 2.26.

(e) Muscle B has the larger maximal isometric force than muscle A, and because of its shorter fibres, muscle B has a smaller maximal speed of shortening than muscle A (assuming the same fibre type distribution in both muscles); see Figure 2.27.

58   *Theoretical models of skeletal muscle*

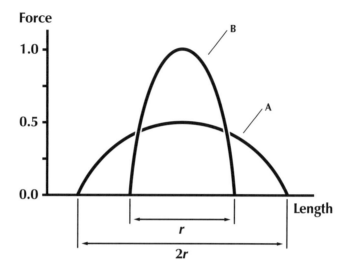

Figure 2.26 Solution to Problem 2.1d.

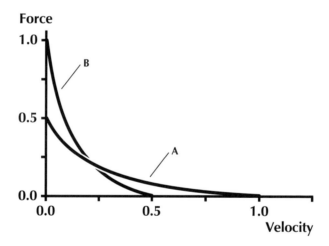

Figure 2.27 Solution to Problem 2.1e.

(f) The required graph (Figure 2.28) follows directly from the force–velocity graph of the two muscles. (power = force · velocity).

2.2 Using Hill's [48] force–velocity equation (equation (2.8)) and assuming that $a/F_0 = b/v_0 = 0.25$ (equation (2.9)), calculate the velocity of shortening, v, relative to the maximal velocity, v_0, at which the power produced by a muscle is maximal.

Modelling skeletal muscle using simple geometric shapes

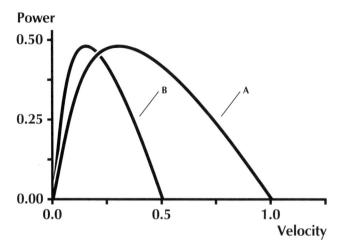

Figure 2.28 Solution to Problem 2.1f.

Solution
The following relations are given:

$$F = \frac{F_0 b - av}{b + v} \quad (1)$$

$$a/F_0 = b/v_0 = 0.25 \quad (2)$$

Muscular power, $P(v)$, is defined as the product of force and velocity ($F \cdot v$). Therefore, for a given force–velocity relation of a muscle, the maximal power as a function of velocity, $P_0(v)$, can be obtained in the following way. By definition,

$$P(v) = F(v) \cdot v \quad (3)$$

and

$$\frac{dP(v)}{dv} = \frac{dF}{dv} v + F(v) \quad (4)$$

Using Hill's force–velocity relation given above (equation (1)), equation (4) may be written as:

$$\frac{dP(v)}{d(v)} = \frac{(F_0 + a)b^2 - a(v + b)^2}{(v + b)^2} \quad (5)$$

Realizing that $dP(v)/dv$ must be zero for $P(v)$ to become maximal, we have:

$$0 = \frac{(F_0 + a)b^2 - a(v + b)^2}{(v + b)^2} \quad (6)$$

solving equation (6) for the velocity, v_m at which maximal power occurs, gives:

$$v_m = b(\sqrt{(F_0/a) + 1} - 1) \qquad (7)$$

solving equation (2) for a and b and substituting into equation (7) gives:

$$v_m = \frac{v_0}{4}(\sqrt{4+1} - 1) \qquad (8)$$

or

$$v_m = 0.31 v_0.$$

The maximal power of the muscle whose force–velocity behaviour is described by equations 1 and 2 occurs at a velocity of shortening that is about 31% of the maximal velocity of shortening.

2.3 Search the literature for the physiological cross-sectional area and fibre length of a human skeletal muscle (e.g. vastus lateralis), assume $a/F_0 = b/v_0 = 0.25$ for all fibres, and then determine the force–length and the force–velocity property of that muscle.

Solution
Literature search [101] reveals that the physiological cross-sectional area (PCSA) and the optimal fibre lengths of the human vastus lateralis are $PCSA = 30 \text{ cm}^2$ and $L_{f0} = 8$ cm.
From equation (2.3), and assuming an average value for the specific force constant of $k = 30 \text{ N/cm}^2$, we obtain that the maximal isometric force at optimal length (F_0) for the vastus lateralis is about:

$$F_0 = 30 \text{ N/cm}^2 \cdot 30 \text{ cm}^2 = 900 \text{ N}$$

Using the values for F_0 calculated above and for L_{f0} found in the literature, the normalized version of the force–length relation described in equation (2.6) for an entire muscle may be used to calculate the actual force–length relation of the vastus lateralis (Figure 2.29).
In order to determine the force–velocity relation, the only remaining unknown is the maximal velocity of shortening, v_0. The magnitude of v_0 depends primarily on two factors: the optimal length of the fibre, L_{f0}, and the fibre type (slow, fast). For slow and fast fibres, v_0 is approximately:

$$v_{0,\text{slow}} \approx 8 \cdot L_{f0}/s \approx 64 \text{ cm/s}$$

$$v_{0,\text{fast}} \approx 16 \cdot L_{f0}/s \approx 128 \text{ cm/s}$$

therefore:

$$a = 0.25 \cdot F_0 = 225 \text{ N}$$

$$b_{\text{slow}} = 0.25 \cdot v_0 = 16 \text{ cm/s}$$

$$b_{\text{fast}} = 0.25 \cdot v_0 = 32 \text{ cm/s}$$

Modelling skeletal muscle using simple geometric shapes 61

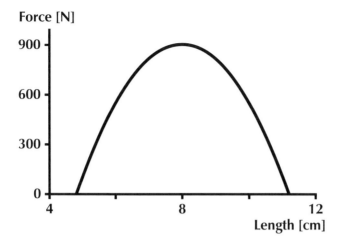

Figure 2.29 Solution to Problem 2.3. Force–length property of the human vastus lateralis muscle.

and so, the force–velocity relation for a slow fibred vastus lateralis becomes

$$F = \frac{F_0 b_s - av}{b_s + v}$$

and for a fast fibred vastus lateralis, it becomes

$$F = \frac{F_0 b_f - av}{b_f + v}$$

where the subscripts f and s refer to fast and slow, respectively.
In actuality, of course, the human vastus lateralis is a mixed-fibred muscle, and therefore, it might be expected that its actual force–velocity properties lie somewhere between the two extreme cases shown in Figure 2.30.

2.4 Consider the force–velocity properties for shortening of two isolated muscle fibres shown in Figure 2.31. What can be concluded based on this graph in terms of:

(a) the cross-sectional area of the two fibres?
(b) the fibre type of the two fibres?
(c) the power production of the two fibres?

Justify your answers.

Solution
(a) Since the maximal isometric forces of fibres A and B are equal, it can be assumed that their cross-sectional areas are about the same.

62   Theoretical models of skeletal muscle

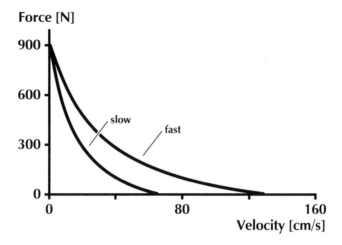

Figure 2.30 Solution to Problem 2.3. Force–velocity property of the human vastus lateralis muscle assuming a purely fast- and a purely slow-twitch fibre composition.

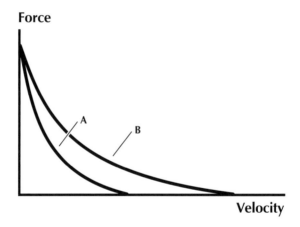

Figure 2.31 Problem formulation for Problem 2.4. Schematic force–velocity relation of two muscle fibres, A and B.

(b) Nothing can be said about the fibre type of the two fibres (i.e. if they are slow- or fast-twitch fibres). Fibre B has the higher maximal speed of shortening. Assuming that the fibres are of equal length, one could safely assume that fibre B was of a 'faster' type than fibre A. However, one could also assume that they are both of the same fibre type. If that were the case, the difference in the maximal speed of shortening would probably be associated with different lengths of the fibres: fibre B being longer than fibre A.

Modelling skeletal muscle using simple geometric shapes

(c) The instantaneous power of fibre B (force × velocity) will be higher than that for fibre A at any speed of fibre shortening, because fibre B has a higher force than fibre A for any given speed of shortening.

2.5 Plot a force–length–velocity relation in a three-dimensional graph, assuming that the relation is governed by the equation.

$$F = \left(\frac{F_0 b - av}{b + v}\right) f(l)$$

and $f(l)$, the force as a function of muscle length, is given by

$$f(l) = -6.25(L/L_0)^2 + 12.5(L/L_0) - 5.25$$

where the maximal isometric force, F_0, is $F_0 = 1000$ N; the maximal velocity of shortening, v_0, is $v_0 = 100$ cm/s; and a and b are given by

$$a = 0.25 F_0,$$
$$b = 0.25 v_0$$

and the optimal length of the muscle, L_0, is $L_0 = 10$ cm.
Note that, for $f(l)$ to be positive, L may vary in the range 6–14 cm. Consider only shortening velocities, i.e., $v = 0$–100 cm/s.

Solution
See Figure 2.32

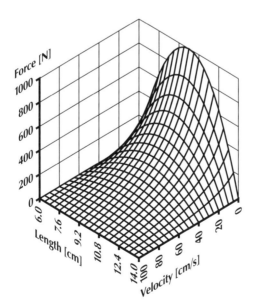

Figure 2.32 Solution to Problem 2.5. Three-dimensional representation of the force length–velocity properties of the muscle defined in Problem 2.5.

REFERENCES

[1] Abbott, B.C. and Aubert, X.M. (1952) The force exerted by active striated muscle during and after change of length. *J. Physiol.* (London), **117**: 77–86.
[2] Abbott, B.C. and Wilkie, D.R. (1953) The relation between velocity of shortening and the tension–length curve of skeletal muscle. *J. Physiol.* (London), **120**: 214–223.
[3] Alexander, R.M. (1984) Elastic energy stores in running vertebrates. *Am. Zool.*, **24**: 85–94.
[4] Alexander, R.M. and Vernon, A. (1975) The dimensions of knee and ankle muscles and the forces they exert. *J. Hum. Mvmt Studies*, **1**: 115–123.
[5] Allen, G.M., Gandevia, S.C. and McKenzie, D.K. (1995) Reliability of measurement of muscle strength and voluntary activation using twitch interpolation. *Muscle & Nerve*, **18**: 593–600.
[6] Allinger, T.L., Epstein, M. and Herzog, W. (1996) Stability of muscle fibers on the descending limb of the force–length relation. A theoretical consideration. *J. Biomech.*, **29**: 627–633.
[7] Amemiya, Y., Iwamoto, H., Kobayashi, T., Sugi, H., Tanaka, H. and Wakabayashi, K. (1988). Time-resolved X-ray diffraction studies on the effect of slow length changes on tetanized frog skeletal muscle. *J. Physiol.* (London), **407**: 231–241.
[8] Archambault, J.M. and Herzog, W. (1997) Strain behaviour and mechanical properties of the cat patellar tendon during submaximal loading. *J. Biomech.* Submitted.
[9] Baskin, R.J. and Paolini, P.J. (1967) Volume change and pressure development in muscle during contraction. *Am. J. Physiol.*, **213**: 1025–1030.
[10] Belanger, A.Y. and McComas, A.J. (1981) Extent of motor unit activation during effort. *J. Appl. Physiol.*, **51**(5): 1131–1135.
[11] Benedict, J.V., Walker, D.B. and Harris, E.H. (1968) Stress–strain characteristics and tensile strength of unembalmed human tendon. *J. Biomech.*, **1**: 53–63.
[12] Benninghof, A. and Rollhäuser, H. (1952) Zur inneren Mechanik des gefiederten Muskels. *Pflügers Arch.*, **254**: 527–548.
[13] Blix, M. (1894) Die Laenge und die Spannung des Muskels. *Skand. Arch. Physiol.*, **5**: 149–206.
[14] Bobbert, M.F. and van Soest, A.J. (1992) Effects of muscle strengthening on vertical jump height: a simulation study. *Med. Sci. Sports Exerc.*, **26**: 1012–1020.
[15] Brooks, J.G., Herzog, W. and Leonard, T.R. (1994) Fiber dynamics of unipennate cat medial gastrocnemius during active shortening. *Proc. 8th Conf. CSB*, 72–73 (abstract).
[16] Butler, D.L., Sheh, M.Y., Stouffer, D.C., Samaranayake, V.A. and Levy, M.S. (1990). Surface strain variation in human patellar tendon and knee cruciate ligaments. *J. Biomech. Engng*, **112**: 38–45.
[17] Cavagna, G.A., Saibene, F.P. and Margaria, R. (1964) Mechanical work in running. *J. Appl. Physiol.*, **19**: 249–256.
[18] Close, R. (1964) Dynamic properties of fast and slow skeletal muscles of the rat during development. *J. Physiol.* (London), **173**: 74–95.
[19] Close, R. (1965) Force:velocity properties of mouse muscle. *Nature*, **206**: 718–719.
[20] Close, R., and Hoh, J.F.Y. (1967) Force:velocity properties of kitten muscles. *J. Physiol.*, **192**: 815–822.
[21] Close, R.I. (1972) Dynamic properties of mammalian skeletal muscles. *Physiol. Rev.*, **52**: 129–197.
[22] Dengler, R., Stein, R.B. and Thomas, C.K. (1988) Axonal conduction velocity and force of single human motor units. *Muscle & Nerve*, **11**: 136–145.

Modelling skeletal muscle using simple geometric shapes

[23] Denny-Brown, D. (1928) On inhibition as a reflex accompaniment of the tendon jerk and other forms of muscular response. *Proc. R. Soc. London B*, **103**: 321–336.
[24] Edman, K.A.P. (1979) The velocity of unloaded shortening and its relation to sarcomere length and isometric force in vertebrate muscle fibres. *J. Physiol.* (London), **291**: 143–159.
[25] Edman, K.A.P., Elzinga, G. and Noble, M.I.M. (1978) Enhancement of mechanical performance by stretch during tetanic contractions of vertebrate skeletal muscle fibres. *J. Physiol.* (London), **281**: 139–155.
[26] Edman, K.A.P. and Reggiani, C. (1983) Length–tension–velocity relationships studied in short consecutive segments of intact muscle fibres in the frog, in *Contactile Mechanisms of Muscle. Mechanics, Energetics and Molecular Models*, Volume II. (eds G.H. Pollack and H. Sugi), Plenum Press, New York, pp. 495–510.
[27] Ettema, G.J.C. and Huijing, P.A. (1989) Properties of the tendinous structures and series elastic component of EDL muscle–tendon complex of the rat. *J. Biomech.*, **22**: 1209–1215.
[28] Faulkner, J.A., Claflin, D.R. and McCully, K.K. (1986) Power output of fast and slow fibers from human skeletal muscles, in *Human Muscle Power* (eds N.L. Jones, N. McCartney, and A.J. McComas), Human Kinetics Publishers, Champaign, IL.
[29] Faulkner, J.A., Jones, D.A., Round, J.M. and Edwards, R.H.T. (1980) Dynamics of energetic processes in human muscle. *Proc. Int. Symp. Exer. Bioenergetics Gas Exchange, Milan* pp. 81–90.
[30] Ford, L.E., Huxley, A.F. and Simmons, R.M. (1981) The relation between stiffness and filament overlap in stimulated frog muscle fibres. *J. Physiol.* (London), **311**: 219–249.
[31] Gans, C. and Bock, W.J. (1965) The functional significance of muscle architecture – a theoretical analysis. *Ergebn. Anat. Entw. Gesch.*, **38**: 115–142.
[32] Gordon, A.M., Huxley, A.F. and Julian, F.J. (1966) The variation in isometric tension with sarcomere length in vertebrate muscle fibres. *J. Physiol.* (London), **184**: 170–192.
[33] Goslow, G.E., Jr. and Van De Graaff, K.M. (1982) Hindlimb joint angle changes and action of the primary ankle extensor muscles during posture and locomotion in the striped skunk (*Mephitis mephitis*). *J. Zool.* (London), **197**: 405–419.
[34] Granzier, H.L.M. and Pollack, G.H. (1989) Effect of active pre-shortening on isometric and isotonic performance of single frog muscle fibres. *J. Physiol.*, **415**: 299–327.
[35] Griffiths, R.I. (1989) The mechanics of the medial gastrocnemius muscle in the freely hopping wallaby (*Thylogale billardierii*). *J. Exp. Biol.*, **147**: 439–456.
[36] Griffiths, R.I. (1991) Shortening of muscle fibres during stretch of the active cat medial gastrocnemius muscle: the role of tendon compliance. *J. Physiol.* (London), **436**: 219–236.
[37] Guimarães, A.C., Herzog, W., Hulliger, M., Zhang, Y.T. and Day, S. (1994) Effects of muscle length on the EMG–force relation of the cat soleus muscle using non-periodic stimulation of ventral root filaments. *J. Exp. Biol.*, **193**: 49–64.
[38] Hales, J.P. and Gandevia, S.C. (1988) Assessment of maximal voluntary contraction with twitch interpolation: An instrument to measure twitch response. *J. Neurosci. Methods*, **25**: 97–102.
[39] Heckman, C.J., Weytjens, J.L.F. and Loeb, G.E. (1992) Effect of velocity and mechanical history on the forces of motor units in the cat medial gastrocnemius muscle. *J. Neurophysiol.*, **68**: 1503–1515.

[40] Herzog, W. (1994) Muscle, *Biomechanics of the Musculo-Skeletal System* (eds B.M. Nigg and W. Herzog), John Wiley & Sons, Chichester, pp. 154–190.
[41] Herzog, W., Guimarães, A.C., Anton, M.G. and Carter-Erdman, K.A. (1991) Moment–length relations of rectus femoris muscles of speed skaters/cyclists and runners. *Med. Sci. Sports Exerc.*, **23**: 1289–1296.
[42] Herzog, W., Kamal, S. and Clarke, H.D. (1992) Myofilament lengths of cat skeletal muscle: Theoretical considerations and functional implications. *J. Biomech.*, **25**: 945–948.
[43] Herzog, W., Leonard, T.R., Renaud, J.M., Wallace, J., Chaki, G. and Bornemisza, S. (1992) Force–length properties and functional demands of cat gastrocnemius, soleus and plantaris muscles. *J. Biomech.*, **25**: 1329–1335.
[44] Herzog, W. and Read, L.J. (1993) Lines of action and moment arms of the major force-carrying structures crossing the human knee joint. *J. Anat.*, **182**: 213–230.
[45] Herzog, W., Read, L.J. and ter Keurs, H.E.D.J. (1991) Experimental determination of force–length relations of intact human gastrocnemius muscles. *Clin. Biomech.*, **6**: 230–238.
[46] Herzog, W. and ter Keurs, H.E.D.J. (1988) Force–length relation of *in vivo* human rectus femoris muscles. *Eur. J. Physiol.*, **411**: 642–647.
[47] Heukelom, B., Stelt, A. and Diegenbach, P.C. (1979) A simple anatomical model of muscle and the effects of internal pressure. *Bull. Math. Biol.*, **41**: 791–802.
[48] Hill, A.V. (1938) The heat of shortening and the dynamic constants of muscle. *Proc. R. Soc. London B*, **126**: 136–195.
[49] Hill, A.V. (1953) The mechanics of active muscle. *Proc. R. Soc. London B*, **141**: 104–117.
[50] Hill, A.V. (1970) *First and Last Experiments in Muscle Mechanics*. Cambridge University Press, Cambridge.
[51] Hoffer, J.A., Caputi, A.A., Pose, I.E. and Griffiths, R.I. (1989) Roles of muscle activity and load on the relationship between muscle spindle length and whole muscle length in the freely walking cat, in *Progress in Brain Research*, (eds J.H.H. Allum and M. Hulliger), Elsevier Science Publishers, Amsterdam, pp. 75–85.
[52] Huxley, A.F. (1957) Muscle structure and theories of contraction. *Prog. Biophys. Biophys. Chem.*, **7**: 255–318.
[53] Huxley, A.F. and Simmons, R.M. (1971) Proposed mechanism of force generation in striated muscle. *Nature*, **233**: 533–538.
[54] Huxley, H.E., Stewart, A., Sosa, H. and Irving, T. (1994) X-ray diffraction measurements of the extensibility of actin and myosin filaments in contracting muscles. *Biophys. J.*, **67**: 2411–2421.
[55] Irving, M. (1995) Give in the filaments. *Nature*, **374**: 14–15.
[56] Julian, F.J. and Morgan, D.L. (1979) Intersarcomere dynamics during fixed-end tetanic contractions of frog muscle fibres. *J. Physiol.* (London), **293**: 365–378.
[57] Julian, F.J. and Sollins, M.R. (1975) Variation of muscle stiffness with force at increasing speeds of shortening. *J. Gen. Physiol.*, **66**: 287–302.
[58] Katz, B. (1939) The relation between force and speed in muscular contraction. *J. Physiol.* (London), **96**: 45–64.
[59] Ker, R.F. (1981) Dynamic tensile properties of sheep plantaris tendon (*Ovis aries*). *J. Exp. Biol.*, **93**: 283–302.
[60] Lieber, R.L. and Baskin, R.J. (1983) Intersarcomere dynamics of single muscle fibres during fixed-end tetani. *J. Gen. Physiol.*, **82**: 347–364.
[61] Lombardi, V. and Piazzesi, G. (1992) Force response in steady lengthening of active single muscle fibres, in *Muscular Contraction* (ed. R.M. Simmons), Cambridge University Press, Cambridge and New York, pp. 237–255.

[62] Luff, A.R. (1981) Dynamic properties of the inferior rectus, extensor digitorum longus, diaphragm and soleus muscle of the mouse. *J. Physiol.* (London), **313**: 161–171.
[63] Lutz, G.J. and Rome, L.C. (1993) Built for jumping: The design of the frog muscular system. *Science*, **263**: 370–372.
[64] MacIntosh, B.R., Herzog, W., Suter, E., Wiley, J.P. and Sokolosky, J. (1993) Human skeletal muscle fiber types and force: Velocity properties: Model and Cybex measurements. *Eur. J. Appl. Physiol.*, **67**: 499–506.
[65] Mai, M.T. and Lieber, R.L. (1990) A model of semitendinosus muscle sarcomere length, knee and hip joint interaction in the frog hindlimb. *J. Biomech.*, **23**: 271–279.
[66] Maréchal, G. and Plaghki, L. (1979) The deficit of the isometric tetanic tension redeveloped after a release of frog muscle at a constant velocity. *J. Gen. Physiol.*, **73**: 453–467.
[67] Merton, P.A. (1954) Voluntary strength and fatigue. *J. Physiol.* (London), **123**: 553–564.
[68] Milner-Brown, H.S. and Brown, W.F. (1976) New methods of estimating the number of motor units in a muscle. *J. Neurol. Neurosurg. Psych.*, **39**: 258–265.
[69] Needham, D.M. (1971) *Machina Carnis*, Cambridge University Press, Cambridge.
[70] Nishizaka, T., Miyata, H., Yoshikawa, H., Ishiwata, S. and Kinosita, K.J. (1995) Unbinding force of a single motor molecule of muscle measured using optical tweezers. *Nature*, **377**: 251–254.
[71] Noyes, F.R., Butler, D.L., Grood, E.S., Zernicke, R.F. and Hefzy, M.S. (1984) Biomechanical analysis of human ligament grafts used in knee-ligament repairs and reconstruction. *J. Bone Joint Surg. B*, **71**: 344–352.
[72] Otten, E. (1988) Concepts and models of functional architecture in skeletal muscle. *Exerc. Sport Sci. Rev.*, **16**: 89–137.
[73] Pandy, M.G. and Zajac, F.E. (1991) Optimal muscular coordination strategies for jumping. *J. Biomech.*, **24**: 1–10.
[74] Pandy, M.G., Zajac, F.E., Sim, E. and Levine, W.S. (1990) An optimal control model for maximum-height human jumping. *J. Biomech.*, **23**: 1185–1198.
[75] Pedotti, A., Krishnan, V.V. and Stark, L. (1978) Optimization of muscle-force sequencing in human locomotion. *Math. Biosci.*, **38**: 57–76.
[76] Perrine, J.J. and Edgerton, V.R. (1978) Muscle force–velocity and power–velocity relationships under isokinetic loading. *Med. Sci. Sports Exerc.*, **10**: 159–166.
[77] Prochazka, A., Stephens, J.A. and Wand, P. (1979) Muscle spindle discharge in normal and obstructed movements. *J. Physiol.* (London), **287**: 57–66.
[78] Rack, P.M.H. and Westbury, D.R. (1969) The effects of length and stimulus rate on tension in the isometric cat soleus muscle. *J. Physiol.* (London), **204**: 443–460.
[79] Ranatunga, K.W. (1981) Temperature-dependence of shortening velocity and rate of isometric tension development in rat skeletal muscle. *J. Physiol.* (London), **329**: 465–483.
[80] Reiners, K., Herdmann, J. and Freund, H.J. (1989) Altered mechanisms of muscular force generation in lower motor neuron disease. *Muscle & Nerve*, **12**: 647–659.
[81] Rigby, B.J., Hirai, N., Spikes, J.D. and Eyring, H. (1959) The mechanical properties of rat tail tendon. *J. Gen. Physiol.*, **43**: 265–283.
[82] Rutherford, O.M., Jones, D.A. and Newham, D.J. (1986) Clinical and experimental application of the percutaneous twitch superimposition technique for the study of human muscle activation. *J. Neurol. Neurosurg. Psych.*, **49**: 1288–1291.
[83] Sacks, R.D. and Roy, R.R. (1982) Architecture of the hind limb muscles of cats: Functional significance. *J. Morph.*, **173**: 185–195.

[84] Scott, S.H. and Loeb, G.E. (1995) Mechanical properties of aponeurosis and tendon of the cat soleus muscle during whole-muscle isometric contraction. *J. Morphol.*, 224: 73–86.
[85] Spector, S.A., Gardiner, P.F., Zernicke, R.F., Roy, R.R. and Edgerton, V.R. (1980) Muscle architecture and force–velocity characteristics of cat soleus and medial gastrocnemius: Implications for motor control. *J. Neurophysiol.*, 44: 951–960.
[86] Stouffer, D.C., Butler, D.L. and Hosny, D. (1985) The relationship between crimp pattern and mechanical response of human patellar tendon-bone units. *J. Biomech. Engng*, 107: 158–165.
[87] Sugi, H. and Tsuchiya, T. (1988) Stiffness changes during enhancement and deficit of isometric force by slow length changes in frog skeletal muscle fibres. *J. Physiol.* (London), 407: 215–229.
[88] Suter, E., Herzog, W. and Huber, A. (1996) Extent of motor unit activation in the quadriceps muscles of healthy subjects. *Muscle & Nerve*, 19: 1046–1048.
[89] Suter, E., Herzog, W., Sokolosky, J., Wiley, J.P. and MacIntosh, B.R. (1993) Muscle fiber type distribution as estimated by Cybex testing and by muscle biopsy. *Med. Sci. Sports Exerc.*, 25: 363–370.
[90] Tajima, Y., Makino, K., Hanguu, T., Wakabayashi, K. and Amemiya, Y. (1994) X-ray evidence for the elongation of thin and thick filaments during isometric contraction of a molluscan smooth muscle. *J. Muscle Res. Cell Motil.*, 15: 659–671.
[91] Ter Keurs, H.E.D.J., Iwazumi, T. and Pollack, G.H. (1978) The sarcomere length–tension relation in skeletal muscle. *J. Gen. Physiol.*, 72: 565–592.
[92] Thorstensson, A., Grimby, G. and Karlsson, J. (1976) Force–velocity relations and fiber composition in human knee extensor muscles. *J. Appl. Physiol.*, 40: 12–15.
[93] Van Leeuwen, J.L. and Spoor, C.W. (1992) On the role of biarticular muscles in human jumping. *J. Biomech.*, 25: 207–209.
[94] Van Soest, A.J., Schwab, A.L., Bobbert, M.F. and van Ingen Schenau, G.J. (1993) The influence of the biarticularity of the gastrocnemius muscle on vertical-jumping achievement. *J. Biomech.*, 26: 1–8.
[95] Wagemans, E. (1989) Relations between architecture and the function of pennate muscles during isometric plantar flexion of the ankle. Doctoral dissertation, University of Leuven, Belgium.
[96] Wagemans, E. and van Leemputte, M. (1989) Some structural parameters of m. gastrocnemius linked in a model. *Proc. 12th Int. Cong. Biomech., Los Angeles*, abstract 316.
[97] Wakabayashi, K., Sugimoto, Y., Tanaka, H., Ueno, Y., Takezawa, Y. and Amemiya, Y. (1994) X-ray diffraction evidence for the extensibility of actin and myosin filaments during muscle contraction. *Biophys. J.*, 67: 2422–2435.
[98] Walker, S.M. and Schrodt, G.R. (1973) I segment lengths and thin filament periods in skeletal muscle fibers of the rhesus monkey and the human. *Anat. Record*, 178: 63–81.
[99] Weber, W. and Weber, E. (1836) *Mechanik der menschlichen Gehwerkzeuge*, W. Fischer Verlag, Göttingen.
[100] Weytjens, J.L.F, (1992) Determinants of cat medial gastrocnemius muscle force during simulated locomotion. PhD thesis, University of Calgary.
[101] Wickiewicz, T.L., Roy, R.R., Powell, P.L. and Edgerton, V.R. (1983) Muscle architecture of the human lower limb. *Clin. Orthop. Related Res.*, 179: 275–283.
[102] Wilkie, D.R, (1968) *Muscle*, Studies in Biology, No. 11, Edward Arnold, London.
[103] Woittiez, R.D., Huijing, P.A., Boom, H.B.K. and Rozendal, R.H. (1984) A three-dimensional muscle model: A quantified relation between form and function of skeletal muscles. *J. Morphol.*, 182: 95–113.

[104] Woittiez, R.D., Huijing, P.A. and Rozendal, R.H. (1983) Influence of muscle architecture on the length–force diagram of mammalian muscle. *Eur. J. Physiol.*, **399**: 275–279.
[105] Woo, S.L.-Y., Johnson, G.A. and Smith, B.A. (1993) Mathematical modeling of ligaments and tendons. *J. Biomech. Engng*, **115**: 468–473.
[106] Yamada, H. (1970) *Strength of Biological Materials*, Williams and Wilkins, Baltimore, MD.
[107] Zernicke, R.F., Butler, D.L., Grood, E.S. and Hefzy, M.S. (1984) Strain topography of human tendon and fascia. *J. Biomech. Engng*, **106**: 177–180.
[108] Zuurbier, C.J., Everard, A.J., van der Wees, P. and Huijing, P.A. (1994) Length–force characteristics of the aponeurosis in the passive and active muscle condition and in the isolated condition. *J. Biomech.*, **27(4)**: 445–453.
[109] Zuurbier, C.J. and Huijing, P.A. (1991) Influence of muscle shortening on the geometry of gastrocnemius medialis muscle of the rat. *Acta Anat.*, **140**: 297–303.
[110] Zuurbier, C.J. and Huijing, P.A. (1992) Influence of muscle geometry on shortening speed of fibre, aponeurosis and muscle. *J. Biomech.*, **25**: 1017–1026.
[111] Zuurbier, C.J. and Huijing, P.A. (1993) Changes in geometry of actively shortening unipennate rat gastrocnemius muscle. *J. Morphol.*, **218**: 167–180.

3
Hill and Huxley type models: biological considerations

3.1 INTRODUCTION

In the previous chapter, we have discussed straight-line models of skeletal muscle. Force production in straight-line models depends on the structural arrangement of fibres within the muscle, and the size and properties of aponeuroses and tendons. Consequently, straight-line models are structural models on the gross morphological level. In this chapter, we would like to consider two very different models, which we call Hill and Huxley type models. **Hill** models of skeletal muscle are named after Archibald Vivian Hill. They are phenomenological in nature and describe the force behaviour of muscles for precisely defined contractile conditions (length, speed). **Huxley** models, or cross-bridge models, are named after Andrew Fielding Huxley. They are structural models which are based on the assumed interaction of thin (actin) and thick (myosin) myofilaments via cross-bridges and the corresponding ideas of force production in muscles (or, better, sarcomeres).

3.2 HILL TYPE MODELS

'It is odd how one's brain fails to work properly when pet theories are involved.' This quote from Hill [17] describes his own feelings when scientific evidence [9] suggested that his viscoelastic theory of muscle contraction was wrong. Gasser and Hill [11] had assumed that the effect of contraction speed on the force exerted by a muscle was caused by an elastic network containing a viscous fluid. A contraction of the muscle with its corresponding change in shape would require the viscous fluid to flow relative to the solid tissue. An increase in the speed of contraction, and therefore an increased rate of change in muscle shape, would cause an increase in the viscous force, which in turn would cause a decrease in the force that could be produced externally by the muscle to do

Hill and Huxley type models 71

mechanical work. Gasser and Hill [11] believed that this viscous model accounted for most of the force–speed relationship observed in contracting skeletal muscle. If this belief were correct, one would expect that the amount of energy lost as heat during contraction should be proportional to the speed of shortening or lengthening as it represents the change in the viscous force with speed. However, when a muscle is stretched at a slow speed (so that it does not 'slip'), the rate of heat production is less than during an isometric contraction [16]. Furthermore, the total rate of energy liberation is greatly increased and decreased for concentric and eccentric contractions, respectively, compared to isometric contractions. If the biochemical energy liberation during contraction was a 'muscle constant' and viscous fluid flow in an elastic network was to account for the changes in force with speed of contraction [11], then the above observations could not have been made. Therefore the viscoelastic model proposed by Gasser and Hill [11] (a viscous fluid in an elastic network) cannot explain the energetic observations made on shortening and lengthening active muscle, and so must be dismissed as a primary mechanism for the force–velocity relationship. This finding, however, does not imply that muscles do not have viscous or viscoelastic elements.

In his famous experiments on the heat of shortening of skeletal muscle which led to the formulation of the Hill model, Hill [16] showed that a muscle produced heat during isometric contractions. When the isometrically contracting muscle was suddenly released under a load which allowed for shortening of the muscle, there was an increase in the rate of heat production which was proportional to the speed of shortening and stopped when the muscle stopped shortening (Figure 3.1a). The total extra heat produced during shortening was proportional to the distance shortened (Figure 3.1b).

When the size of the muscle was accounted for and the stimulation was constant (typically supramaximal), the shortening heat (H) could be expressed as $H = ax$, where x is the distance shortened during the contraction, and a represents a constant proportionality factor (in units of force). The value of a depends on the size, or more precisely the physiological cross-sectional area, of the muscle, and further depends on the level of activation. Hill [17] showed that the value a/P_0 (where P_0 is the maximal isometric force) is reasonably constant (≈ 0.25). This result can be understood by realizing that P_0 (like a) depends on the physiological cross-sectional area and the level of activation (for P_0, the level of activation is maximal, by definition) of the muscle.

During a shortening contraction, a muscle produces extra heat (i.e. heat exceeding that observed during isometric contractions) and mechanical work. Since the extra heat (or shortening heat) is equal to ax, and the work is equal to Px (where P is the force of the muscle), the total energy in excess of that produced during isometric contractions becomes $(P + a)x$. The rate of extra energy liberation becomes $(P + a)\,dx/dt = (P + a)v$, where v is the speed of shortening.

Hill [16] showed that the rate of extra energy liberation during shortening was

Theoretical models of skeletal muscle

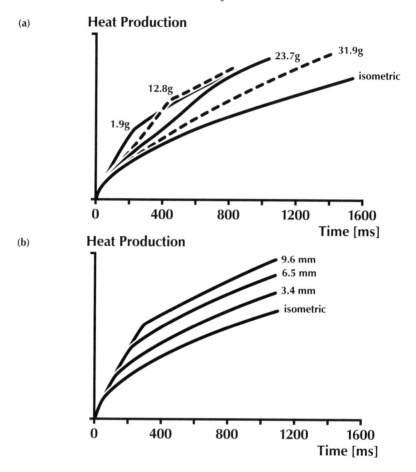

Figure 3.1 Heat production as a function of time during isotonic shortening of tetanized frog skeletal muscle at 0°C: (a) shortening a constant distance using different loads; (b) shortening different distances using a constant load. (Adapted from [16].)

inversely proportional to the load P applied to muscles in afterloaded shortening experiments. By definition, the rate of extra energy liberation is zero during isometric contractions, i.e. when $P = P_0$. Therefore

$$(P + a)v = b(P_0 - P) \tag{3.1}$$

where b is a constant (in units of speed) which defines the absolute rate of energy liberation.

Hill [16] rewrote equation (3.1) as

$$(P + a)(v + b) = (P_0 + a)b = \text{constant} \tag{3.2}$$

Hill and Huxley type models 73

Another way of writing Hill's equation is

$$P = \frac{P_0 b - av}{b + v} \quad (3.3)$$

Hill's equation may be verified by measuring the force (P) for different speeds (v) of shortening, or by measuring the shortening speed for different loads exerted on the muscle. Note that this process does not require any heat measurements. Equations (3.1)–(3.3) describe the loss of force with increasing speeds of shortening for a maximally stimulated muscle at (or near) optimal length. They represent a rectangular hyperbola with asymptotes of $P = -a$ and $v = -b$ (Figure 3.2).

Based on the experimental work available in 1938, Hill [16] concluded that active muscle contains an undamped elastic element and a damped element in series with the undamped elastic element. The initial idea that viscosity was responsible for the damping properties of active skeletal muscles was soon abandoned for a variety of reasons: (1) viscosity did not fit into the experimental observation on lengthening muscle; (2) in order to account for the differences in passive and active muscle properties, viscosity would have had to change dramatically upon activation of the muscle; and (3) Fenn had found that shortening contractions produced more total energy than isometric contractions and suggested that the basis for the apparent viscosity of muscles must lie in the rates of energy yielding chemical reactions rather than the release of elastic energy [9].

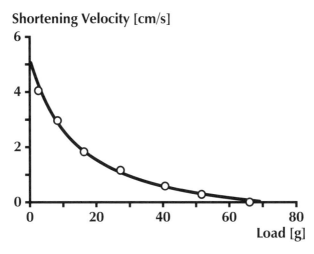

Figure 3.2 Force–velocity relation of tetanized frog striated muscle at 0°C. The circles represent the mean of two experimental observations. The line corresponds to the equation $(P + 14.35)(v + 1.03) = 87.6$; see equation (3.2). (Adapted from Hill [16].)

74 Theoretical models of skeletal muscle

Reviewing all the experimental evidence, Hill [16] concluded that active skeletal muscle is composed primarily of a contractile element in series with a purely elastic element (Figure 3.3). The properties of the contractile element were governed by equation (3.2) which was derived by Hill [16] based on his experimental observations on the heat of shortening in frog skeletal muscle. In order to account for the passive forces observed in stretched skeletal muscles, the basic Hill model (Figure 3.3) is typically supplemented with an undamped elastic spring in parallel with the contractile element (Figure 3.4a) or in parallel with the contractile and series elastic element (Figure 3.4b).

Although Hill type models may be considered purely phenomenological, when using these models, the properties of the contractile and elastic elements need to be known. However, when attempting to determine these properties experimentally, it would be useful to associate the theoretical elements of the model with a biological structure. The association of a theoretical with a biological element is not as trivial as it might appear at first glance. For example, let us consider the series elastic element shown in both parts of Figure 3.4. Many researchers who use Hill type models associate the series elastic element with the tendon. Tendon elastic properties can be determined readily; therefore, if the tendon truly represented the series elastic element, deriving the constitutive properties of the series elastic element would be easy. However, it is well accepted that the aponeurosis of a muscle contributes to the series elasticity [3, 29, 30]. Furthermore, it has

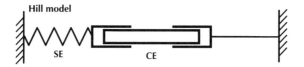

Figure 3.3 Hill model of skeletal muscle with a contractile element (CE) which obeys the characteristic equation (3.3), and an elastic element in series (SE) with the CE.

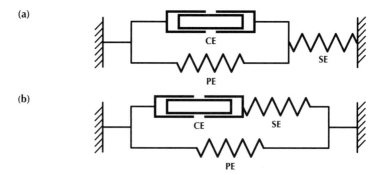

Figure 3.4 Hill models with (a) a parallel elastic (PE) component in parallel with the contractile element (CE), and (b) a parallel elastic component in parallel with the series elastic (SE) and the contractile element.

been shown convincingly that much of the series elasticity of skeletal muscle resides in the cross-bridges [10], i.e. in the contractile portion of Hill type models. And, to complicate things further, the myofilaments which for the past four decades were assumed to be essentially rigid, are now shown to provide as much as 50% of the sarcomere compliance, and therefore should be considered when deriving the properties of the series elastic elements [12, 15, 22, 25, 32]. This example, in part, illustrates one of the basic difficulties associated with Hill type models: the division of the forces between the parallel elastic and the contractile element, and the separation of elongations between the series elastic and the contractile element, are rather arbitrary.

A further limitation of Hill type models based on Hill's [16] characteristic equation is that the characteristic equation is only valid for restricted contractile conditions: maximal activation, shortening contraction, at or near optimal length. Needless to say, these conditions are rarely encountered during voluntary movements in everyday life. The force–velocity behaviour of submaximally contracting muscle is not well understood; also the force–velocity behaviour at lengths other than optimal is not completely understood, and neither are the force–velocity properties for lengthening muscle, specifically in voluntary movements in which maximal activation appears to be limited and force production deviates considerably from lengthening contractions performed in artificially stimulated muscle [33].

A conceptual limitation of Hill type models is that history-dependent force behaviour of skeletal muscle cannot be predicted, unless artificially introduced. Specifically, Hill's characteristic equation implicitly assumes that the relation between force, velocity, and length is unique, but it is not. For example, when shortening at a constant speed to a given final length, the force at the final length will depend directly on the magnitude of the shortening distance [15] (Figure 3.5), a feature which is not captured in Hill type models.

Hill models describe the approximate behaviour of muscles for certain contractile conditions. They do not provide insight into the mechanisms of force production. However, despite these limitations, the Hill model has been (and is) used more frequently in biomechanical models of musculo-skeletal systems than any other muscle model. Its success is associated with the (mathematical) simplicity of the model and the qualitatively correct predictions for a variety of contractile conditions. It is likely that the Hill model, which was derived in its basic form in 1938, will continue to play a major role in muscle modelling for years to come.

3.3 HUXLEY (CROSS-BRIDGE) TYPE MODELS

When concerned with the biophysical events of muscular force production, as well as aspects of muscle energetics, the model of choice for the past 40 years has been the Huxley type or cross-bridge model. Cross-bridge models focus on the contractile element. They integrate the known structural properties of skeletal muscle with the known mechanical and biochemical aspects of contraction.

76   *Theoretical models of skeletal muscle*

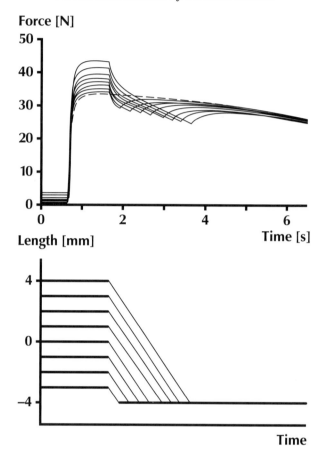

Figure 3.5 Force depression following shortening contractions in cat soleus muscle. The dashed line in the force–time graph represents the isometric reference force. The remaining (solid) force–time traces show the isometric force following shortening to the reference length. Note the decrease in force at the end of shortening with increasing shortening distances.

The cross-bridge theory of muscular force production was first described in detail by Andrew Fielding Huxley [18]. It must be considered the earliest attempt at a unifying theory of muscle contraction – unifying in the sense that it accounted simultaneously for the mechanical, thermal, chemical, and structural changes which were known to occur during contraction. Although other theories of muscular force production were subsequently proposed [23, 28], none of them became universally accepted, primarily because these alternative theories were unable to predict and account for as many observations as accurately and comprehensively as the cross-bridge theory did. Therefore, to this day, the cross-bridge theory is the accepted paradigm of muscular force production, and although many changes have been introduced to the model, the basic theory proposed in 1957 is still recognized.

In contrast to the previously discussed models of skeletal muscle (the straight-line and Hill type models), the cross-bridge model allows for the quantification of heat production and energetics during contraction, and is based on structural observations of the molecular composition of skeletal muscle. The cross-bridge model is structural in nature, whereas Hill type models are phenomenological. Of course, many of the aspects of the cross-bridge model are also phenomenological, because the exact molecular events leading to force production and contraction in skeletal muscles are not known, and therefore were modelled so as to obey observed phenomena.

The cross-bridge model is accepted and used virtually unanimously in biophysical research; in contrast, it is rarely used, with few exceptions [35], in biomechanical applications. The primary reasons for ignoring the cross-bridge model in biomechanics are probably associated with the mathematical complexity of the model formulation and the idea that it represents sarcomere rather than fibre or muscle behaviour. However, there are mathematical formulations of the cross-bridge model which are computationally efficient [35], and exposure to these models may stimulate the use of cross-bridge models in biomechanical research.

In its most simplistic form, the cross-bridge model may be described as follows. There are side projections (cross-bridges) arising from the thick (myosin) filaments which attach to specialized sites on the thin (actin) filaments, and when attached, the cross-bridges pull the thin past the thick filaments, thereby producing force and contraction. In the following, the cross-bridge theory as proposed originally in 1957 is introduced.

3.3.1 The 1957 formulation of the cross-bridge theory

Before 1954, most theories of muscular contraction were based on the idea that shortening and force production were the result of some kind of folding or coiling of the myofilaments (particularly the thick filaments) at specialized sites. However, in 1954 H.E. Huxley and Hansen [21] as well as A.F. Huxley and Niedergerke [19] demonstrated that muscle shortening was not associated with an appreciable amount of myofilament shortening, and therefore postulated that muscle shortening is probably caused by a sliding of the thin past the thick myofilaments (the **sliding filament theory**). The mechanism whereby this myofilament sliding is produced was proposed by A.F. Huxley [18], and is referred to as the **cross-bridge theory**.

In the cross-bridge theory [18], it was assumed that thick filaments had side pieces which were connected via elastic springs to the thick filament. The side piece with its connection point M (Figure 3.6) was thought to oscillate about its equilibrium position (O) because of thermal agitation. M was assumed to attach to specialized sites (A) on the thin filament, if M came into the vicinity of A. The

78 Theoretical models of skeletal muscle

combination of M-sites with A-sites was thought to occur spontaneously and was restricted to occur asymmetrically only on one side of O so that a combination of the M- and A-sites would cause a force (because of the tension in the elastic element constraining the side piece M) and movement which tended to shorten the sarcomere. Attachment and detachment were thought to be governed by rate functions f and g, respectively, and f and g were modelled to be linear functions of the distance from A, the active site on the thin filament, to the equilibrium position, O, of the side piece (denoted x; see Figures 3.6 and 3.7). Since the combination of an M- with an A-site was taken to occur spontaneously, breaking the M–A connection had to be associated with an active, energy-requiring process.

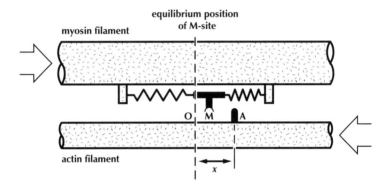

Figure 3.6 Schematic illustration of the 1957 model of the cross-bridge theory. (Adapted from Huxley [18].)

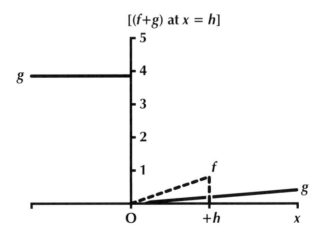

Figure 3.7 Rate functions for the formation, f, and the breaking, g, of cross-bridge links between the thick (myosin) and the thin (actin) myofilaments as a function of x, the distance from the active site on the thin filament to the equilibrium position of the cross-bridge. (Adapted from [18].)

The energy for this process was assumed to come from splitting a high-energy phosphate compound.

For force production to occur smoothly, it was assumed that there were a number of M- and A-sites for possible combination of the thick and thin filaments, which were staggered relative to one another so that different combination sites would come into contact at different relative displacements of the two myofilaments. The M- and A-sites were further assumed to be so far apart that events at one site would not influence events at another site.

The cross-bridge theory and its energetics are assumed to be associated with defined structures. The M-sites are represented by the S1 subfragment of the myosin protein (the cross-bridge, Figure 1.3); the A-sites are the attachment sites on the actin near the troponin (Figure 1.5) and the high-energy phosphate supplying the energy for detachment of the cross-bridges is associated with ATP. Typically, it is assumed that one ATP molecule is hydrolysed per full cross-bridge cycle.

Since a thick myofilament in mammalian skeletal muscle is about 1600 nm in length and contains cross-bridges along its entire length except for about the middle 160 nm, one-half of the thick filament contains about 50 (720 nm / 14.3 nm) pairs of side pieces offset by 180°, and each side piece is thought to contain two cross-bridge heads for possible attachment on the thin filament. Since neighbouring cross-bridge pairs are thought to be offset by 60° (Figure 1.4), there are about 16 (720 nm / 42.9 nm) side pieces available on each thick filament for interaction with a given thin filament.

In order to test the cross-bridge model of muscular contraction, Huxley [18] compared the predictions of his theory (which were formulated in precise mathematical terms (see Chapter 4) with the experimental results obtained by Hill [16] on frog striated muscle during tetanic stimulation at 0°C. Huxley [18] found good agreement between the normalized force–velocity relation of Hill [16] and his own theoretical predictions (Figure 3.8).

When comparing the predictions of the theory to the properties of stimulated muscle which is forcibly stretched, several observations were made. Katz [24] found that the slope of the force–velocity curve for slow lengthening was about six times greater than the corresponding slope for slow shortening. Huxley's [18] theory also predicted this asymmetry in the force–velocity curve about the isometric point, with the slopes differing by a factor of 4.33. Katz [24] further found that the force produced during rapid lengthening of a stimulated muscle was about 1.8 times the isometric force. Using the rate functions given by Huxley [18], the force for increasing speeds of lengthening approaches asymptotically a value of 5.33 times the isometric force. This value is too large.

Similarly, Huxley's [18] theory does not predict well the heat production of a muscle that is stretched. From the theory, it is predicted that the rate of liberation of heat increases linearly with the speed of lengthening, a prediction which vastly overestimates the heat production in lengthening muscle [1, 2]. However, Huxley [18] points out that the discrepancy between experiment and theory

80 Theoretical models of skeletal muscle

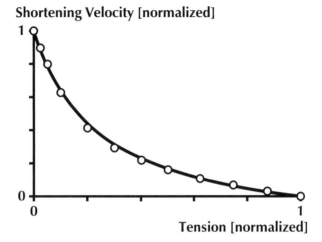

Figure 3.8 Comparison of the force–velocity relationship obtained using Hill's characteristic equation (with $a/P_0 = 0.25$) [16] (solid curve), and obtained by Huxley [18] (dots) based on the cross-bridge model (Adapted from [18].)

could be eliminated quite readily by assuming that during lengthening the cross-bridge connections were broken mechanically rather than released via ATP splitting. This assumption has been implemented in various models recently to account for experimental observations made during concentric [4] and eccentric contractions.

3.3.2 The 1971 formulation of the cross-bridge theory

Another characteristic of muscular contraction which cannot be predicted adequately with the 1957 theory is the force transients following a stepwise length change. When a muscle is shortened rapidly, the force drops virtually simultaneously with the length change and then recovers quickly (Figure 3.9). Two force parameters were defined by Huxley and Simmons [20] for describing these fast force transients: they are referred to as T_1 and T_2. T_1 is defined as the minimum force achieved during the rapid shortening; T_2 is the force at the end of the quick recovery phase (Figure 3.9). T_1 becomes progressively smaller with increasing release distances, and was assumed to be linearly related to the release distance (Figure 3.10). The T_1 versus length step curve was assumed to represent the undamped elasticity of the contractile machinery. T_2 is always larger than T_1, indicating a force recovery within milliseconds of the length step (Figure 3.10).

In the 1957 theory, a cross-bridge was either attached or detached. When a fully activated muscle was shortened rapidly, many cross-bridges would detach during the shortening step, and force recovery was dependent on the rate of cross-bridge attachment. The rate function for attachment, however, was too slow to account

Hill and Huxley type models 81

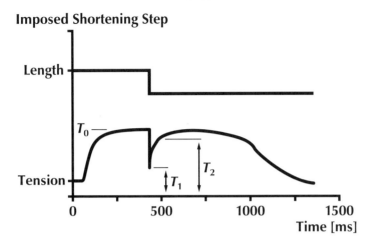

Figure 3.9 Definitions of T_1 and T_2. T_1 is the minimal force value obtained during a rapid release of a muscle; T_2 is the force value achieved following the quick recovery period.

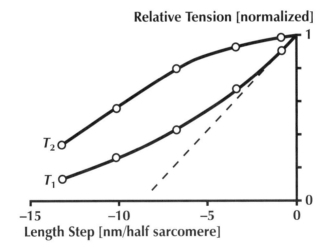

Figure 3.10 T_1 and T_2 as a function of the length step (in nanometres per half sarcomere). The dashed line represents the estimated T_1-curve after correcting for the amount of force recovery which happens during the rapid release. (Adapted from [20].)

for the quick force recovery. An easy way to remedy this limitation would be to increase the attachment rate of cross-bridges [27]. However, models with substantially increased rate functions for attachment could not predict Hill's [16] force–velocity relation as well as the 1957 model, and they could not fit the thermal data observed experimentally during shortening contractions [34].

82 Theoretical models of skeletal muscle

In order to account for the force transients following a stepwise length change and to not lose the good predictive power of the 1957 model, Huxley and Simmons [20] introduced the concept of different states of attachment for the cross-bridge, thereby allowing the cross-bridge to perform work (while it is attached) in a small number of steps. Going from one stable attachment to the next was associated with a progressively lower potential energy. Furthermore, Huxley and Simmons [20] assumed that there is an undamped elastic element within each cross-bridge which allows the cross-bridge to go from one stable attachment state to the next without a corresponding relative displacement of the thick and thin filaments. A diagrammatic representation of the 1971 cross-bridge model is shown in Figure 3.11.

The force transients during a rapid length change are now explained as follows. If a muscle is released infinitely fast, there will be no rotation of the cross-bridge head (Figure 3.12a, b). Therefore, the drop in force observed during the length step (T_1) corresponds to the force–elongation property of the undamped elastic element within the cross-bridge. Since it had been argued that the relationship between the T_1-value and the distance of the length step was virtually linear (the experimentally observed nonlinearity was associated with the beginning of the quick recovery during the large length steps), the cross-bridge elasticity was assumed to be linear as well (2.3×10^{-4} N/m [20]). Once the infinitely fast length step has been completed, the quick recovery of force is possible because of a rotation of the cross-bridge head from a position of high to a position of low potential energy, thereby stretching the elastic link in the cross-bridge, and so increasing the cross-bridge force (Figure 3.12c).

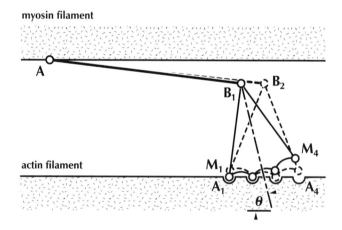

Figure 3.11 Schematic representation of the assumed interaction between the thick and thin filaments according to the 1971 cross-bridge theory [20]. The cross-bridge head is attached to the thick myofilament via an elastic spring. The cross-bridge head can rotate, and so produce different amounts of tension in the elastic link AB without relative movements of the myofilaments. (Adapted from [20].)

Hill and Huxley type models 83

Huxley and Simmons [20] discussed a cross-bridge model with three stable, attached states and derived equations for a system containing two stable states. Many further models with a variety of stable states have been proposed [7, 8] but the basic ideas of these models can all be traced to the 1971 cross-bridge model [20].

The cross-bridge model, as discussed here, has dominated our thinking on muscular contraction for the past 40 years. It does not account for all observed

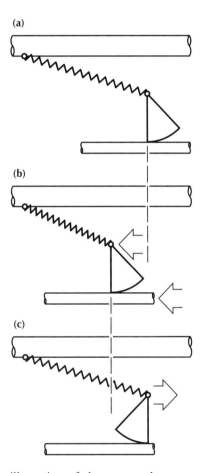

Figure 3.12 Schematic illustration of the presumed events associated with a rapid release and the following quick recovery of force. (a) The cross-bridge head is in its initial position and the elastic link is stretched. (b) A rapid release has occurred. The cross-bridge head is in the same orientation as in (a) but the elastic link has shortened because of the relative movement of the myofilaments. The cross-bridge force (carried by the elastic link) is smaller in (b) than in (a). (c) The cross-bridge head rotates to a position of lower potential energy, thereby stretching the elastic link and increasing the cross-bridge force without any myofilament movement.

84 Theoretical models of skeletal muscle

phenomena; in fact one might argue that it neglects some very basic phenomena, such as the long-lasting, history-dependent force production of muscle following stretch or shortening [1, 5, 6, 13, 26, 31]. Therefore, it is probable that the cross-bridge model may be revised or replaced in the near future. However, at present, it represents the paradigm of choice and it will require strong evidence and convincing theory to replace it. There does not appear to be a serious alternative at the time of writing (1997).

In the following chapter, the mathematical treatment of the cross-bridge theory will be introduced.

PROBLEMS

3.1 Although Hill's 1938 manuscript [16] is typically cited as the classic work on the force–velocity relationship, it was not the first to show that the speed of shortening was inversely related to the maximal active force-producing ability of muscle. Find works before 1938 which give a quantitative account of the force–velocity relationship of skeletal muscle.

3.2 Describe two alternative theories of muscular contraction that have been published in the scientific literature and that are conceptually different from the cross-bridge model. What are the limitations of these alternative theories? Why, therefore, have they not become accepted paradigms for explaining the mechanism of force production in skeletal muscle?

3.3 What are some of the limitations of the Hill type models and the 1957 and 1971 cross-bridge models?

3.4 It is typically assumed that one cross-bridge cycle is associated with the hydrolysation of one ATP molecule. How much energy is freed up by hydrolysing one ATP molecule?

3.5 Hill type muscle models are considered phenomenological models; cross-bridge models are typically interpreted as structural models. However, many aspects of muscular force production are not known in detail, therefore many aspects of the cross-bridge model were derived to account for experimentally observed phenomena. Name some of the phenomenological aspects of the 1971 cross-bridge model, and of cross-bridge models today.

REFERENCES

[1] Abbott, B.C. and Aubert, X.M. (1952) The force exerted by active striated muscle during and after change of length. *J. Physiol.* (London), **117**: 77–86.

[2] Abbott, B.C. and Wilkie, D.R. (1953) The relation between velocity of shortening and the tension–length curve of skeletal muscle. *J. Physiol.* (London), **120**: 214–223.

[3] Baratta, R. and Solomonow, M. (1991) Dynamic performance of a load-moving skeletal muscle. *J. Appl. Physiol.*, **71**: 749–757.

[4] Cooke, R., White, H. and Pate, E. (1994) A model of the release of myosin heads from actin in rapidly contracting muscle fibers. *Biophys. J.*, **66**: 778–788.

[5] Edman, K.A.P., Caputo, C. and Lou, F. (1993) Depression of tetanic force induced by loaded shortening of frog muscle fibres. *J. Physiol.* (London), **466**: 535–552.
[6] Edman, K.A.P., Elzinga, G. and Noble, M.I.M. (1978) Enhancement of mechanical performance by stretch during tetanic contractions of vertebrate skeletal muscle fibres. *J. Physiol.* (London), **281**: 139–155.
[7] Eisenberg, E. and Greene, L.E. (1980) The relation of muscle biochemistry to muscle physiology. *Annual Rev. Physiol.*, **42**: 293–309.
[8] Eisenberg, E., Hill, T.L. and Chen, Y.D. (1980) Cross-bridge model of muscle contraction: quantitative analysis. *Biophys. J.*, **29**: 195–227.
[9] Fenn, W.O. (1923) A quantitative comparison between the energy liberated and the work performed by the isolated sartorius of the frog. *J. Physiol.* (London), **58**: 175–205.
[10] Ford, L.E., Huxley, A.F. and Simmons, R.M. (1981) The relation between stiffness and filament overlap in stimulated frog muscle fibres. *J. Physiol.* (London), **311**: 219–249.
[11] Gasser, H.S. and Hill, A.V. (1924) The dynamics of muscular contraction. *Proc. R. Soc. London B*, **96**: 398–437.
[12] Goldman, Y.E. and Huxley, A.F. (1994) Actin compliance: are you pulling my chain? *Biophys. J.*, **67**: 2131–2136.
[13] Granzier, H.L.M. and Pollack, G.H. (1989) Effect of active pre-shortening on isometric and isotonic performance of single frog muscle fibres. *J. Physiol.* (London), **415**: 299–327.
[14] Herzog, W. and Leonard, T.R. (1997) Depression of cat soleus force following isokinetic shortening. *J. Biomech.*, **30**, 865–872.
[15] Higuchi, H., Yanagida, T. and Goldman, Y.E. (1995) Compliance of thin filaments in skinned fibers of rabbit skeletal muscle. *Biophys. J.*, **69**: 1000–1010.
[16] Hill, A.V. (1938) The heat of shortening and the dynamic constants of muscle. *Proc. R. Soc. London B*, **126**: 136–195.
[17] Hill, A.V. (1970) *First and Last Experiments in Muscle Mechanics*. Cambridge University Press, Cambridge.
[18] Huxley, A.F. (1957) Muscle structure and theories of contraction. *Prog. Biophys. Biophys. Chem.*, **7**: 255–318.
[19] Huxley, A.F. and Niedergerke, R. (1954) Structural changes in muscle during contraction. Interference microscopy of living muscle fibres. *Nature*, **173**: 971–973.
[20] Huxley, A.F. and Simmons, R.M. (1971) Proposed mechanism of force generation in striated muscle. *Nature*, **233**: 533–538.
[21] Huxley, H.E. and Hansen, J. (1954) Changes in cross-striations of muscle during contraction and stretch and their structural implications. *Nature*, **173**: 973–976.
[22] Huxley, H.E., Stewart, A., Sosa, H. and Irving, T. (1994) X-ray diffraction measurements of the extensibility of actin and myosin filaments in contracting muscles. *Biophys. J.*, **67**: 2411–2421.
[23] Iwazumi, T. (1979) A new field theory of muscle contraction, in *Cross-bridge Mechanism in Muscle Contraction* (eds H. Sugi and G.H. Pollack), University of Tokyo Press, Tokyo, pp. 611–632.
[24] Katz, B. (1939) The relation between force and speed in muscular contraction. *J. Physiol.* (London), **96**: 45–64.
[25] Kojima, H., Ishijima, A. and Yanagida, T. (1994) Direct measurement of stiffness of single actin filaments with and without tropomyosin by in vitro nanomanipulation. *Proc. Natl. Acad. Sci. USA*, **91**: 12 962–12 966.
[26] Maréchal, G. and Plaghki, L. (1979) The deficit of the isometric tetanic tension redeveloped after a release of frog muscle at a constant velocity. *J. Gen. Physiol.*, **73**: 453–467.

[27] Podolsky, R.J. (1960) Kinetics of muscular contraction: the approach to the steady state. *Nature*, **188**: 666–668.
[28] Pollack, G.H. (1990) *Muscles and Molecules: Uncovering the Principles of Biological Motion*, Ebner and Sons, Seattle.
[29] Roeleveld, K., Baratta, R.V., Solomonow, M., van Soest, A.G. and Huijing, P.A. (1993). Role of tendon properties on the dynamic performance of different isometric muscles. *J. Appl. Physiol.*, **74**: 1348–1355.
[30] Scott, S.H. and Loeb, G.E. (1995) Mechanical properties of aponeurosis and tendon of the cat soleus muscle during whole-muscle isometric contraction. *J. Morphol.*, **224**: 73–86.
[31] Sugi, H. and Tsuchiya, T. (1988) Stiffness changes during enhancement and deficit of isometric force by slow length changes in frog skeletal muscle fibres. *J. Physiol.* (London), **407**: 215–229.
[32] Wakabayashi, K., Sugimoto, Y., Tanaka, H., Ueno, Y., Takezawa, Y. and Amemiya, Y. (1994) X-ray diffraction evidence for the extensibility of actin and myosin filaments during muscle contraction. *Biophys. J.*, **67**: 2422–2435.
[33] Westing, S.H., Seger, J.Y. and Thorstensson, A. (1990) Effects of electrical stimulation on eccentric and concentric torque-velocity relationships during knee extension in man. *Acta Physiol. Scand.*, **140**: 17–22.
[34] Woledge, R.C., Curtin, N.A. and Homsher, E. (1985) *Energetic Aspects of Muscle Contraction*, Academic Press, London.
[35] Zahalak, G.I. (1981) A distribution-moment approximation for kinetic theories of muscular contraction. *Math. Biosci.*, **55**: 89–114.

4
Rheological and structural models: mathematical considerations

4.1 RHEOLOGICAL MODELS

4.1.1 Introduction

Rheological models are often used as phenomenological tools to describe the mechanical response of **anelastic materials**, such as skeletal muscle. These models consist of various arrangements of rheological elements. A rheological element can be seen as a 'black box' with a definite deterministic relation between a force history and an elongation history. In other words, if a force history $f(t)$ is used as an input (Figure 4.1), then the displacement $u(t)$ is obtained uniquely by some mathematical operation characterizing the element, and vice versa. In many cases it is convenient to make the force–displacement history relation depend on one or more parameters, e.g. temperature or neural activation.

The most common rheological elements are the **linear elastic element** or **spring** (Figure 4.2), and the **linear viscoelastic element** or **dashpot** (Figure 4.3). The spring is characterized by an **unstretched length**, L_0, and a **stiffness**

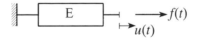

Figure 4.1 A rheological element.

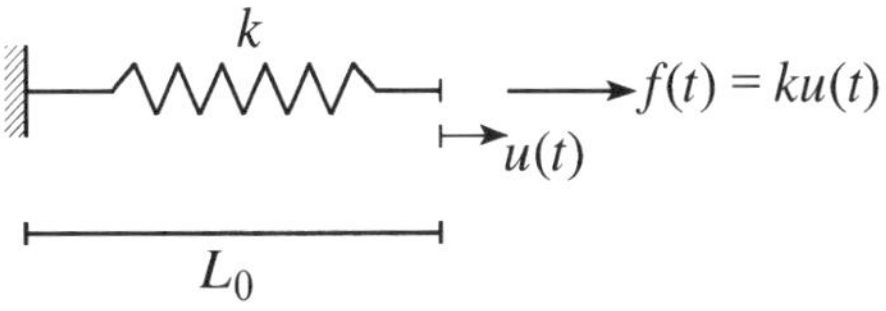

Figure 4.2 A linear spring.

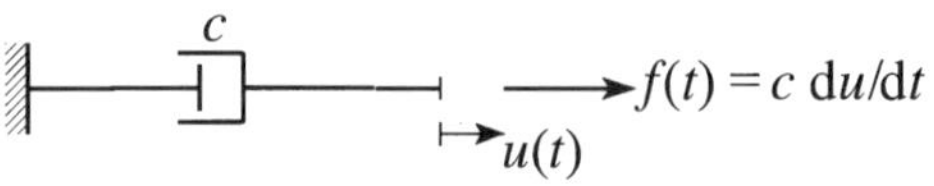

88   *Theoretical models of skeletal muscle*

$$\xrightarrow{c}\quad f(t) = c\, du/dt$$
$$\mapsto u(t)$$

Figure 4.3 A linear dashpot.

constant, k. The force at any instant is proportional to the displacement measured from the unstretched configuration, the constant of proportionality being k. The dashpot is characterized by a **viscous constant**, c, whereby the force at any instant is proportional to the instantaneous value of the velocity $\dot{u} = du/dt$. These are just two examples of rheological elements. Other examples are nonlinear springs and dashpots, dry-friction elements, thermoelastic elements, and purely contractile elements.

Rheological elements can be combined in two ways: in **series** or in **parallel**. In a series arrangement (Figure 4.4), the force is always the same in both elements, while the elongations sum to the total elongation of the combined element. In a parallel arrangement (Figure 4.5), the elongations are the same and the forces are added. The variety of responses that can be obtained by combining linear springs and dashpots in series and in parallel is truly amazing. Three classical models so obtained are: the **Maxwell body**, the **Voigt body**, and the **Kelvin body** (see Problems 4.1–4.3). In each of these examples the force–elongation relation can be completely described by means of a linear first-order ordinary differential equation of the form

$$Af + B\dot{f} = Cu + D\dot{u} \qquad (4.1)$$

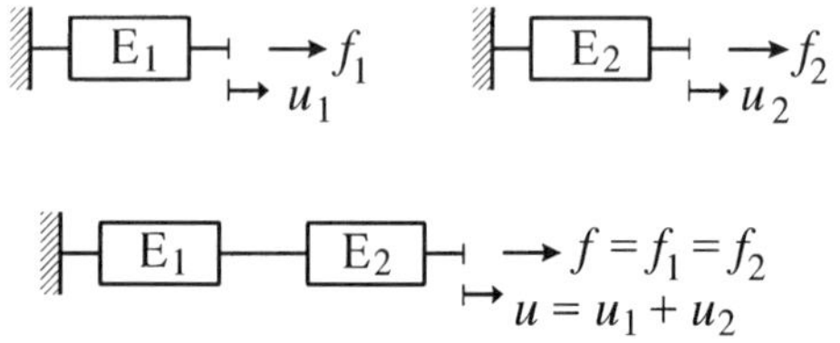

$$\boxed{E_1} \to f_1 \qquad \boxed{E_2} \to f_2$$
$$\mapsto u_1 \qquad \qquad \mapsto u_2$$

$$\boxed{E_1}\!-\!\boxed{E_2} \to f = f_1 = f_2$$
$$\mapsto u = u_1 + u_2$$

Figure 4.4 Two elements in series.

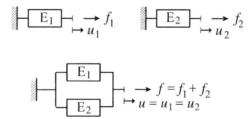

Figure 4.5 Two elements in parallel.

Rheological and structural models

where A, B, C, D are constants. The linearity of the equation stems directly from the assumed linearity of the springs and dashpots. As for the first-order character (namely, the order of the highest derivative appearing in the equation), a word of warning is required. If more first-order elements are added, so that extra 'internal' degrees of freedom become available, the elimination of the corresponding internal displacements or forces may yield a higher-order differential relation (see Problem 4.5).

4.1.2 Hill's three-element model

In his landmark 1938 paper [4], A.V. Hill concluded the presentation and interpretation of his experimental results with the assertion that skeletal muscle may be seen as a 'two-component system, consisting of an undamped purely elastic element in series with a contractile element governed by the characteristic equation $(P + a)(v + b) = $ const.' A standard modification of this idea is the introduction of an extra elastic element in parallel with Hill's two-element combination (Figure 4.6). The contractile element (CE) is a device that introduces the force–velocity relation 'through the back door', as it were, rather than obtaining it as a consequence of the combined behaviour of simpler elements. By the same back-door technique, Hill's CE (originally conceived to operate at the plateau of the force–length relation) can be generalized to include the full force–length response. Such a generalized CE, to be used in the following model, is governed by an all-or-nothing activation parameter. In the inactive state, the CE cannot sustain any force, and its length can be adjusted at will. In the active state, on the other hand, the behaviour of the CE may be described by the equation:

$$f_{CE} = f_{CE}(\dot{w}, l_0) \qquad (4.2)$$

providing the force as a function of the speed of elongation $\dot{w}$, and the length l_0 of the CE *at the moment of activation*. This function can be related to empirical data as follows. Let the experimental force–velocity relation be symbolically expressed as

$$f_{exp} = f_{exp}(\dot{w}, F_{max}) \qquad (4.3)$$

where F_{max} is the maximal isometric force at the plateau of the force–length relation. For lengths other than optimal, the formula should be scaled down by

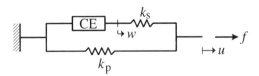

Figure 4.6 Hill's model, consisting of a contractile element (CE), a series-elastic (k_s) and a parallel-elastic element (k_p).

Theoretical models of skeletal muscle

the ratio $F(l_0)/F_{max}$, where $F(l_0)$ is the active force–length relation evaluated at l_0. This scaling down can be done in at least two different manners: either by requiring that

$$f_{CE}(\dot{w}, l_0) = f_{exp}(\dot{w}, F(l_0)) \tag{4.4}$$

or, alternatively, by

$$f_{CE} = \frac{F(l_0)}{F_{max}} f_{exp}(\dot{w}, F_{max}) \tag{4.5}$$

As a viable example of the former option one may suggest formulae of the following type:

$$f_{CE} = \begin{cases} 0 & \text{for } \dot{w} \leq -F(l_0)\dfrac{b}{a} \\[6pt] \dfrac{F(l_0)b + a\dot{w}}{-\dot{w} + b} & \text{for } -F(l_0)\dfrac{b}{a} < \dot{w} \leq 0 \\[6pt] 1.5 F(l_0) - 0.5 \dfrac{F(l_0)b' - a'\dot{w}}{\dot{w} + b'} & \text{for } 0 < \dot{w} \leq F(l_0)\dfrac{b'}{a'} \\[6pt] 1.5 F(l_0) & \text{for } F(l_0)\dfrac{b'}{a'} < \dot{w} \end{cases} \tag{4.6}$$

where a, b, a', b' are constants. In the shortening range the equation is a direct rewriting of Hill's law (except that the speed $\dot{w}$ is considered positive in elongation). The overall appearance of equation (4.6) is plotted in Figure 4.7.

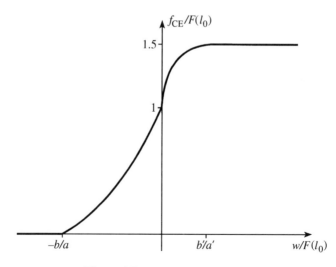

Figure 4.7 Force–velocity relation.

Rheological and structural models

To derive the equations governing this three-element model, let us agree to measure displacements u from the unique resting (inactive, force-free) state corresponding to the unstretched length, L_p, of the parallel element and the initial length, l_i of the inactive contractile element. The unstretched length of the series elastic element is, therefore, $L_s = L_p - l_i$. Let u_0 be the value of u at the instant in which the activation is applied. At that moment, the elongation of the contractile element must also be u_0, since the force in the series spring (and hence its elongation) will vanish up to the instant just before the activation takes place. It follows that the length of the contractile element at the moment of activation must be:

$$l_0 = L_p + u_0 - L_s = l_i + u_0 \tag{4.7}$$

We conclude that the response of the three-element model is governed by the equations

$$f = k_p u + k_s(u - w) \tag{4.8}$$

and

$$k_s(u - w) = f_{CE}[\dot{w}, l_0] \tag{4.9}$$

with l_0 given by equation (4.7) and w denoting the elongation of the contractile element with respect to its length at the resting state of the system. Though not necessary, it is possible to eliminate the internal degree of freedom w by reading it off equation (4.8) and introducing the result in equation (4.9) to obtain the first-order nonlinear differential equation:

$$f - k_p u = f_{CE}\left[\left(1 + \frac{k_p}{k_s}\right)\dot{u} - \frac{\dot{f}}{k_s}, l_0\right] \tag{4.10}$$

If the applied force $f(t)$ is given, equation (4.10) may be used to solve for $u(t)$, and vice versa. Naturally, appropriate initial conditions are to be specified in each case.

4.1.3 Critique and possible extensions of Hill's model

There is a subtle point hidden in our proposed equations (4.8) and (4.9) which needs further elucidation. For definiteness, we will assume that the function f_{CE} is given by equation (4.6). Suppose that the muscle is fully activated at an initial length l_0 and then further stretched at a very slow speed (i.e. quasi-statically). According to equation (4.9), we obtain (with $\dot{w} = 0$):

$$w = u - \frac{F(l_0)}{k_s} \tag{4.11}$$

and, by equation (4.8),

$$f = k_p u + F(l_0) \tag{4.12}$$

92 Theoretical models of skeletal muscle

In other words, the force increases linearly with the displacement, always with a positive stiffness equal to k_p. This is represented schematically in Figure 4.8. Notice that this fact remains true regardless of whether the initial length happens to fall on the ascending or on the descending limb of the force–length relation. This is an important feature, since it means that this model is **unconditionally stable** at all lengths. This feature stands in sharp contrast with models in which the force is made to depend on the *instantaneous length* directly, erroneously using the isometric force–length relation for non-isometric conditions and thereby necessarily leading to instability in the (softening) descending limb.[1] In actual fact, softening behaviour of muscle has not been observed experimentally. Having said this, we must emphasize that the instantaneous positive stiffness experimentally observed upon active stretching is not constant, but rather depends on the initial length of activation l_0. This feature can be incorporated in the model either directly (see Problem 4.8), or by introducing yet another parallel elastic element, but one which functions much as an **elastic rack** which is engaged only upon activation (cf. [3], where the model has the added advantage that Hill's law need not be assumed *ab initio*, but can be obtained as a consequence of the mutual interaction of more or less standard rheological elements).

A final word needs to be added regarding temporal changes in neural activation. Any complete muscle model should allow for such variations, but definitive experimental data are not available. Consider, for example, the following question. If a cycle were effected consisting of: (i) isometric activation at length l_0;

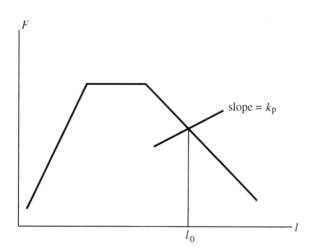

Figure 4.8 Force response to a stretching upon full isometric activation.

---

[1] An elastic material is said to exhibit softening behaviour if the slope of the stress–strain curve is negative at a point. This negative instantaneous stiffness gives rise to instability in the sense that, for a small perturbation of an equilibrium state, the restoring elastic forces will not be sufficient to overcome the disturbing effect of the external forces, and the perturbation will tend to grow.

Rheological and structural models 93

(ii) stretch; (iii) partial deactivation; (iv) shortening to length l_0; and (v) reactivation to a 100% level– what would the resultant force be? Moreover, if the steps were changed to: (i′) partial isometric activation at length l_0; (ii′) stretch; (iii′) full activation; and (iv′) shortening to length l_0 – would the result be the same as before? This example introduces the possibility of **non-commutativity**. These and other **memory effects** are still awaiting a proper experimental understanding. In the meantime, the modeller must resort to provisional solutions which at least do not violate the basic laws of physics. In our three-element Hill model, for instance, one could assume that the muscle 'remembers' the length l_0 at which it was activated for as long as any non-zero activation remains. As soon, and only as soon, as the activation disappears, that length is 'forgotten' and a new initial length will be determined upon the reappearance of any level of activation. For partial activation, equation (4.10) could be modified to read

$$f - k_p u = \beta f_{CE}\left[\left(1 + \frac{k_p}{k_s}\right)\dot{u} - \frac{\dot{f}}{k_s}, l_0\right] \quad (4.13)$$

where β is an activation parameter ranging between 0 (no activation) and 1 (full activation). This, of course, is not the only possible way to tackle the activation issue.

4.1.4 Example: protocols of stretch

In order better to appreciate the merits and shortcomings of a Hill type model such as the one proposed, we will now evaluate the behaviour of the three-element model outlined above for the protocols sketched in Figure 4.9. The resting state is at 125% of optimal length for the contractile element. The constants to be used in equations (4.6) will be assumed to have the values: $a = 10$ N, $b = 40$ mm/s, $a' = 10$ N, $b' = 30$ mm/s, $k_s = 10$ N/mm, $k_p = 1$ N/mm. The maximal isometric force will be taken as $F_{max} = 45$ N, and the force–length relation will be assumed approximated by the parabola:

$$F(l) = (-0.772r^2 + 1.544r - 0.494)\frac{F_{max}}{0.278} \quad (4.14)$$

where r is the ratio between length at activation and optimal length. This parabola attains positive values within the range $0.4 < r < 1.6$. Although of reasonable magnitude, these data are used for illustration purposes only.

A computer code was written which implements the solution of the differential equation (4.13) by means of a forward finite-difference scheme: the time derivative of the force is approximated in terms of the known value at time i and the unknown value at time $i + 1$ as

$$\dot{f} \cong \frac{f_{i+1} - f_i}{h}$$

94 Theoretical models of skeletal muscle

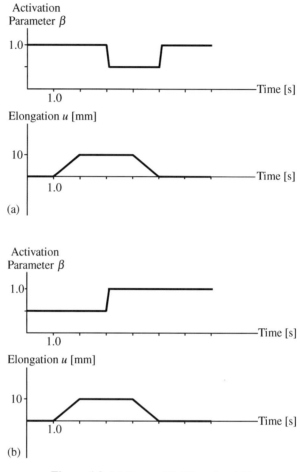

Figure 4.9 (a) Protocol I; (b) protocol II.

where h is a small time increment. To test accuracy, the program was run with different values of h. The results, in terms of force versus time, are plotted in Figure 4.10. The following points are worthy of mention. First, although the activation is applied instantaneously, the velocity dependence of Hill's law brings about a viscous-like effect, so that the force grows gradually, rather than instantaneously, to its full isometric value. Secondly, the general appearance of the force response curves is strikingly similar to the experimental observations (cf. Figure 2.17), particularly in terms of the exponential-like growth towards increased values upon stretching (and the decay counterpart for shortening). Thirdly, because of the unsophisticated way in which partial activation has been implemented, the results are 'commutative' in the sense discussed above. Finally, although the initial length of the CE lies on the descending limb of the force–length relation, the solution is stable.

Rheological and structural models 95

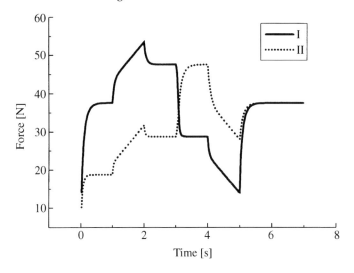

Figure 4.10 Force response.

4.1.5 Example: effect of speed and lingering time

Using the same data as for the previous example, we wish to evaluate the force response for a fully activated muscle subjected to a stretch followed by a shortening of the same magnitude. We wish to study the response for two different speeds and for four different amounts of lingering before the application of the shortening. The different protocols are shown in Figure 4.11.

Using the same computer code as for the previous example, the force responses shown in Figure 4.12 were obtained. Again, the response curves are typical of those obtained in the laboratory (cf. Figure 2.16), although some secondary effects may not have been completely incorporated in the capabilities of the model (incomplete force recovery, possibly unequal depth of valleys, etc). Some of these and other 'missing' effects could be included in the model by modifying the response functionals of one or more of the three constitutive elements, or by adding more elements. Nevertheless, it is quite remarkable that the basic qualitative behaviour of a large number of muscular activities can be captured by a relatively simple combination of a Hill-like contractile element and two linear springs. We emphasize again that a key point in this rather successful exploitation of an old idea is the concept of using the force–length equation not as an instantaneous relation between force and length but as a point of departure for the determination of the maximum available isometric force. The muscle, then, 'remembers' the length at which it was first activated until the activation disappears. Generalizations of this idea are certainly possible, particularly with regard to partial deactivation and reactivation. On the other hand, since a phenomenological model does not pretend to explain but only to describe a certain

96 Theoretical models of skeletal muscle

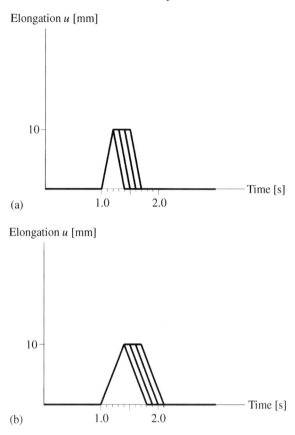

Figure 4.11 (a) Faster stretch; (b) slower stretch.

range of phenomena, the desirability of such generalizations will depend on the degree of detail sought in practical applications.

4.1.6 Softening and stability

In this subsection we will produce a purely mechanical analogue of the force–length relation. We will concentrate on the simulation of the descending limb only, since it appears to exhibit softening behaviour and yet the muscle is stable against small perturbations around an equilibrium state. Is it possible for a material to exhibit a global softening behaviour while being infinitesimally stable? The answer is a qualified 'yes', as the following example shows. The existence of a microstructure, not accounted for in a usual load–deflection experiment, is the essential stabilizing factor (cf. [1]).

Figure 4.13 shows a (two-dimensional) brush consisting of n equal bristles a

Rheological and structural models

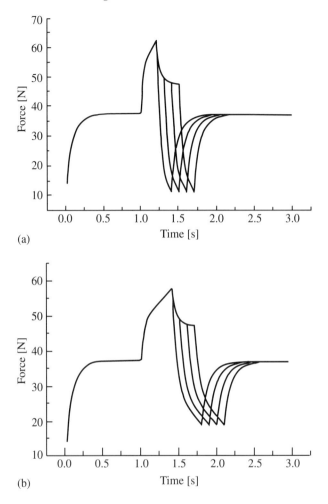

Figure 4.12 Force response to (a) faster stretch, (b) slower stretch.

distance s apart. As two such brushes are placed opposite to each other, so that their bristles interpenetrate, a stiffness against relative lateral movement develops. If the brushes are placed inside a 'black box", in such a way that only their handles are visible from the outside (Figure 4.14), and if the relative lateral displacement of these handles is controlled and the force measured, an apparently softening behaviour is observed.

In order better to describe and understand this phenomenon we start by considering the load–deflection (F–Δ) response of one bristle pair. As shown in Figure 4.15, depending on the initial depth of penetration p, the bristles will be in contact until a relative displacement d takes place. The horizontal force neces-

98   *Theoretical models of skeletal muscle*

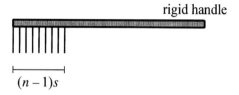

Figure 4.13 A brush with n bristles.

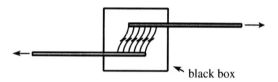

Figure 4.14 Two brushes in mesh.

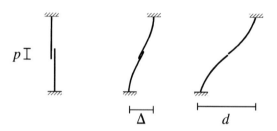

Figure 4.15 Progressive interaction up to impending disengagement.

sary to achieve this state depends on the bending stiffness of the bristles as well as on the interpenetration p. Assuming, for the sake of simplicity, a linear force–displacement behaviour, the load–deflection diagram for one bristle pair is shown in Figure 4.16.

We assume that $s < d$, namely, that the bristles are close enough so that, when pairs disengage, contact is immediately established with the neighbouring bristles. When this disengagement takes place, then, the force on the re-engaged bristles (or, more precisely, its horizontal component) is instantaneously reduced to the value:

$$g = \left(1 - \frac{s}{d}\right)f \qquad (4.15)$$

as can be seen from Figure 4.17.

For two brushes with all n bristle pairs initially engaged, the force–deflection behaviour, for a steadily increasing deflection, is shown in Figure 4.18. The

Rheological and structural models

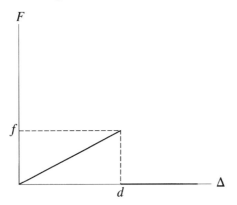

Figure 4.16 Force–displacement relation for one bristle pair.

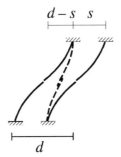

Figure 4.17 Disengagement and re-engagement.

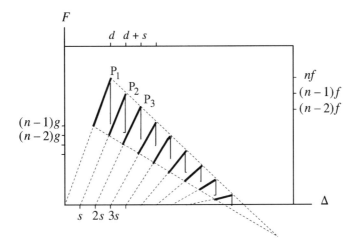

Figure 4.18 Overall force–displacement response.

peaks $(P_1, P_2, P_3, \ldots)$ lie on the straight line with equation

$$F_p = \left(n - \frac{\Delta - d}{s}\right)f \tag{4.16}$$

and the valleys, on the line

$$F_v = \left(n - 1 - \frac{\Delta - d}{s}\right)g \tag{4.17}$$

both having a negative slope. But, except at the peaks themselves, the stiffness of the load–deflection line is always positive. In a black-box experiment, only a finite number of points lying between the line of peaks and the line of valleys will be captured, thus showing an apparent softening behaviour (Figure 4.19).

Returning now to the amount of interpenetration, p, let us assume that the box carries a small lever which permits the vertical distance between the handles of the brushes to be adjusted. When $p < 0$, the brushes are disengaged and no force can be transmitted. As p becomes positive and larger, the maximum force that can be produced increases, as also does the stiffness (and the range of attachment d). We see, therefore, that p is the analogue of the activation parameter β.

This mechanism does not entirely 'solve' the instability problem. Indeed, if a gradually increasing force is applied, the response will climb one of the inclines (depending on the number of initially engaged pairs) and, upon reaching its peak, the system will exhibit instability. Nevertheless, this example opens the possibility for the development of models which, by incorporating extra microscopic

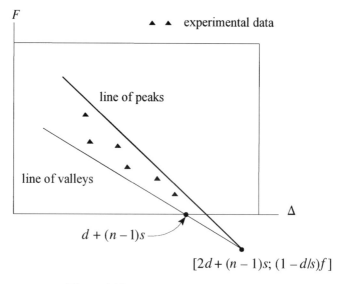

Figure 4.19 Possible experimental results.

Rheological and structural models

structure, may represent a variety of observed phenomena whose subtlety escapes the conventional rheological models.

4.2 STRUCTURAL MODELS

4.2.1 Introduction

In a structural model, as opposed to a phenomenological one, the behaviour of the muscle at the macroscopic level is explained in terms of a detailed description of the events taking place at the 'microscopic' domain. It is clear, however, that the distinction between both types of model is rather arbitrary. For, on the one hand, if a rheological (phenomenological) model agrees with experiment, it may well reveal the existence of an actual microscopic mechanism; and, on the other hand, a structural model will, of necessity, stop at some level of analysis (the cell, the molecule, the atom?) and, therefore, must be phenomenological to a certain degree. In this regard, a point of caution must be borne in mind: if the structural model is based on too low a level, its integration into a macroscopic entity may not lead to the observed global behaviour. As an example, consider the attempt to explain the deformation of a steel beam in terms of the laws of quantum mechanics! The genius of a structural model is, therefore, to find the golden mean in the form of the ideal combination of levels. In the case of skeletal muscle it can be said that the cross-bridge model, proposed by A.F. Huxley in his famous 1957 mémoire [5], has struck that balance and maintained its freshness and vigour over the last four decades. The model is based on a plausible interpretation of well-known observations available at that time (some of which were obtained with the optical microscope) involving the changes of striations of skeletal muscle upon contraction. It should be pointed out that there is still some controversy as to whether or not the cross-bridge hypothesis used by Huxley is indeed the right explanation of the observed facts [8]. Nevertheless, Huxley's model has the merit of providing not only a plausible qualitative interpretation of the sliding filament phenomenon, but also a quantitative description in terms of a manageable differential equation whose coefficients are experimentally traceable to the underlying chemistry.

4.2.2 Huxley's cross-bridge model

The thick (myosin) and thin (actin) filaments can slide parallel to each other in a one-dimensional rectilinear motion. In the original version of the model [5], Huxley assumed that the myosin filament is endowed with 'side-pieces which can slide along the main backbone of the filament, the extent of the movement being limited by an elastic connection'. In Figure 4.20, these moving attachments are denoted by M, and the total spring constant by k. The M-pieces may establish a chemico-mechanical bridge between the filaments by attaching themselves with specific sites A fixed along the adjacent actin filaments. These

102 Theoretical models of skeletal muscle

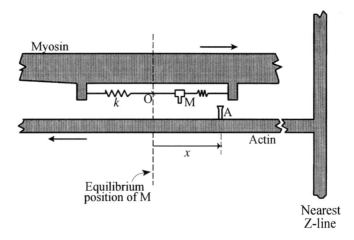

Figure 4.20 Huxley's model.

attachments can also be broken by a chemical reaction 'requiring energy to be supplied by metabolic sources'. Although, as a result of thermal agitation, the unattached sliding element M oscillates, it is assumed that the probability of attachment per unit time[2] is governed by the distance x between the equilibrium (average) position O and the potential attachment site A. This 'probability' distribution is denoted $f(x)$. Conversely, the probability (per unit time) that, once formed, the connection will be broken, is also a function of x, denoted $g(x)$.

If we consider, as Huxley did, a large number of identical M–A pairs (i.e. pairs having at each instant of time one and the same value of x), the proportion $n(t)$ of attached pairs will be a function of time alone. We are interested in obtaining a formula for the rate of change of $n(t)$. By definition of $f(x)$ and $g(x)$, we may write

$$\frac{dn}{dt} = (1-n)f(x) - ng(x) \qquad (4.18)$$

which will henceforth be referred to as **Huxley's equation**. Note that for a state of dynamic equilibrium, that is, when $dn/dt = 0$, we must have, according to Huxley's equation, the following value for the proportion of attached pairs:

$$n_{eq} = \frac{f(x)}{f(x) + g(x)} \qquad (4.19)$$

---

[2] The concept of 'probability per unit time' is meaningless. What is meant here is that the chemical reactions are assumed to be governed by first-order kinetics, whereby the reaction constants are loosely called probabilities per unit time. In a first-order reaction, the number of molecules of a substance which are reacting per unit time is proportional to the instantaneous value of the concentration of that substance, without being affected by the presence of other molecules.

Rheological and structural models 103

This is expected on intuitive grounds: the proportion of attached cross-bridges at equilibrium is governed by the probability of attachment.

In order to solve equation (4.18) for $n(t)$, we must specify the global relative motion $x = x(t)$, and the initial condition $n_0 = n(0)$. It is sometimes convenient to provide, instead of $x(t)$, the global relative sliding velocity $v = v(t)$, in which case $x(t)$ can be obtained by integration as

$$x(t) = x(0) + \int_0^t v(\tau)\, d\tau \qquad (4.20)$$

We note that, by the sign convention implied in Figure 4.20, the arrows representing myofilament sliding indicate a motion with negative v, which we take to mean sarcomere contraction. If, as usually assumed, v is to be positive in contraction, one should change the plus sign in the right-hand side of equation (4.20) to a minus sign.

So far, we have assumed all M–A pairs to be in the same state, namely, to have the same distance $x(t)$. More realistically, though, at any given instant of time, and for a large ensemble of pairs, $x(t)$ would be distributed almost randomly (i.e. uniformly) over the range $[-0.5 l_a, 0.5 l_a]$, where l_a is the typical distance between actin sites.[3] In that case, we must talk of a distribution function $n(x, t)$ 'per unit length', namely, such that the product $n(x, t)\, dx$ represents, at time t, the proportion of attached cross-bridges whose distance from the (nearest) actin site lies between x and $x + dx$.[4] By the uniformity assumption, the proportion of unattached pairs in the same interval $(x, x + dx)$ is given by

$$\left[\frac{1}{l_a} - n(x, t) \right] dx \qquad (4.21)$$

Again, we are interested in obtaining the rate of change of $n(x, t)$ with time, as seen by an observer fixed at an actin site. It should be noted that this is not the same as the partial derivative $\partial n(x, t)/\partial t$, since, by definition, the partial derivative is calculated while holding x fixed. To take into consideration the fact that, for a fixed material position along the actin filament, x too varies with time (after all, the filaments *are* moving relative to each other!), we calculate the material rate of change with time by the chain rule of differentiation as

$$\frac{Dn}{Dt} = \frac{\partial n}{\partial t} + \frac{\partial n}{\partial x} v \qquad (4.22)$$

where, as before, we retain the natural sign convention for v (negative in contraction). A careful analysis, in which the assumption of filament rigidity plays a crucial role, confirms that the governing differential equation is now

---

[3] Actin sites beyond this range disappear from the picture or are replaced by the next entrant site.
[4] The units of $n(x, t)$ are, therefore, reciprocal length, as opposed to Huxley's $n(t)$, which is dimensionless. Some authors [9] prefer to define a dimensionless $n(x, t)$ by multiplying by l_a. We do not adopt such a procedure.

104 Theoretical models of skeletal muscle

$$\frac{\partial n}{\partial t} + \frac{\partial n}{\partial x} v = \left(\frac{1}{l_a} - n\right) f(x) - n g(x) \qquad (4.23)$$

where equation (4.21) was taken into consideration *vis-à-vis* its counterpart in equation (4.18).[5] We shall call the first-order partial differential equation (4.23) the **Huxley–Zahalak equation**, since Zahalak [9] seems to have been the first to formulate it in this manner. The global motion in terms, say, of the relative sliding velocity $v(t)$ is assumed to be given a priori. To solve an equation of this type (i.e. a first-order partial differential equation) one has to specify the values n_0 of the unknown variable n on an initial curve (or initial manifold), most commonly the line $t = 0$.[6] At any given point P of the initial manifold (Figure 4.21), there passes a unique **characteristic curve**, $x = \gamma_p(t)$, obtained as the solution of the ordinary differential equation:

$$\frac{d\gamma_p(t)}{dt} = v(t) \qquad (4.24)$$

with the initial condition

$$\gamma_p(0) = x_p \qquad (4.25)$$

Unlike the general case shown in Figure 4.21, in our particular case the characteristic curves are all 'parallel' to each other, since at any time t the slope $v(t)$ is independent of x. A careful look at the Huxley–Zahalak equation reveals that its left-hand side represents the time derivative of the function $n(\gamma_p(t), t)$, where $\gamma_p(t)$ is the unique characteristic curve passing through the point (x, t). (This follows directly by the chain rule of differentiation and by equation (4.24), which gives the slope of the characteristic curve through the point (x, t).) In other words, the integration of the partial differential equation (4.23) boils down to the integration

---

[5] If the filaments were not assumed rigid, one would have to incorporate an additional convected term containing the velocity gradient along the filament [7]. Briefly, the argument leading to equation (4.23) is based on the integral equation

$$\frac{D}{Dt} \int_{a(t)}^{b(t)} n(x, t) \, dx = \int_{a(t)}^{b(t)} \left(\frac{1}{l_a} - n\right) f(x) dx - \int_{a(t)}^{b(t)} n g(x) \, dx$$

where $a(t)$ and $b(t)$ are any two values of x. The subtle point to understand is that in order for this equation to make physical sense, we must ensure that the *same* M–A pairs are 'caught' in the left-hand-side integral as time goes on. If the filaments are rigid, this can be achieved by 'following' the pairs by means of the instantaneous value of the sliding speed $v(t)$. If the filaments are not rigid, the speed itself will be a function of a lengthwise coordinate, y, which must be introduced explicitly in the formulation.

[6] This should be contrasted with the analogous initial value problem for a first-order *ordinary* differential equation, where the initial value of the unknown function must be specified for one point only. In the case of the partial differential equation, the unknown function depends also on the spatial variable x, so that an initial value must be specified *for each x*, giving thus rise to the notion of the initial curve (or, in the general case of many variables, the initial manifold).

Rheological and structural models

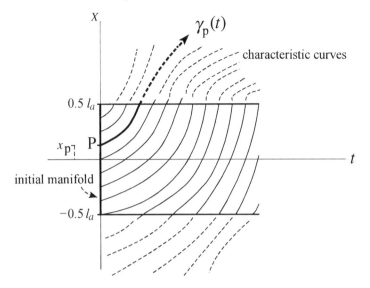

Figure 4.21 Characteristic curves.

of a Huxley type ordinary differential equation along the characteristic curve through each point P of the initial manifold, namely

$$\frac{dn_p(t)}{dt} = \left(\frac{1}{l_a} - n_p(t)\right) f(\gamma_\pi(t)) - n_p(t) g(\gamma_p(t)) \qquad (4.26)$$

with the initial condition

$$n_p(0) = n_0(P) \qquad (4.27)$$

A similar mathematical treatment applies to more general partial differential equations of the first order in two independent variables. For a complete analysis, the reader is referred to [2].

4.2.3 The attachment/detachment probability distributions

As already pointed out, the distributions $f(x)$ and $g(x)$ represent the rate constants of the chemical reactions associated with attachment and detachment of M–A pairs, respectively. They are expressed in units of reciprocal time. For the model to yield the desired results, it is assumed that $f(x)$ vanishes if A is to the left of M, whereas, for the same condition, $g(x)$ attains a very large constant value g_0. When A is to the right of M, it is assumed that both $f(x)$ and $g(x)$ increase linearly, $f(x)$ being truncated at a value $x = h$ (much smaller than $0.5 l_a$) representing the range of bonding ability (Figure 4.22). An explanation for such odd behaviour of the probability distributions may be found within a more

106   *Theoretical models of skeletal muscle*

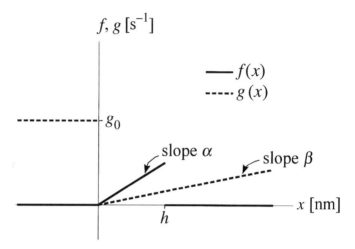

Figure 4.22 Huxley's 'probability' distributions.

detailed geometric description of the cross-bridges and the molecular structure, which was not available to Huxley in 1957.

4.2.4 Macroscopic quantities and the cross-bridge model

We wish to calculate the total internal force in a muscle with a parallel fibre arrangement (Figure 4.23) on the basis of the cross-bridge model. Since half-sarcomeres in series must produce the same force, we need only consider the sum of the contributions of all the half-sarcomeres cut by a normal cross-section. Let A represent the current area of the cross section, m the number of M-sites per unit volume, and s the current average sarcomere length. Then, the number of sites contained in all the half-sarcomeres affected by the cross-section is equal to $mAs/2$. The average force per site is obtained by calculating the weighted

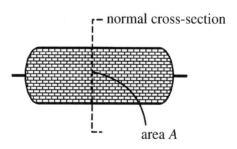

Figure 4.23 A parallel-fibre muscle.

Rheological and structural models

average of the forces in the individual springs:

$$f_{\text{ave}}(t) = \int_{-\infty}^{\infty} kxn(x,t)\, dx \qquad (4.28)$$

Here the weighting function is, naturally, the proportion $n(x, t)$ of attached cross-bridges per unit length. The total force is

$$F(t) = \frac{mAs}{2} \int_{-\infty}^{\infty} kxn(x,t)\, dx \qquad (4.29)$$

obtained by multiplying the average force per site by the number of sites involved in the cross-section.

Let the muscle be given a constant negative (i.e. contractile) velocity V. This results in a sliding sarcomere velocity:

$$v = \frac{V}{n_s} \qquad (4.30)$$

where n_s is the number of half-sarcomeres in one muscle length. Since v in this case is a constant, integration of equation (4.24) yields the characteristic curves as the straight lines:

$$\gamma_p(t) = x_p + vt, \qquad v < 0 \qquad (4.31)$$

A crucial point at this juncture is that, instead of assuming initial conditions at $t = 0$, Huxley effectively moves the initial manifold to the line $x = h$ (Figure 4.24), and specifies that $n \equiv 0$ on this line.[7] At first sight it may seem surprising that 'initial' conditions could be specified on a line other than t = constant. But the line $x = h$ is a perfectly valid initial manifold, since it satisfies the mathematical restriction of not being anywhere tangent to a characteristic line.[8]

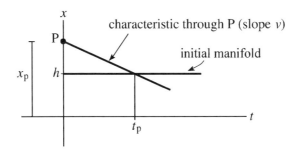

Figure 4.24 Huxley's initial manifold.

---

[7] Huxley [5] does not explain his procedure in these terms, since he does not deal with a partial differential equation. Huxley's procedure, on the other hand, would break down if the sliding velocity were not assumed constant in time.

[8] The first-order partial differential equation can be seen as providing information about the rate of change of the unknown function in the local characteristic direction. Therefore, if the initial data happen to lie along a characteristic line, the equation is not able to provide information necessary to 'come out' of this line.

108 Theoretical models of skeletal muscle

The time of intersection of the characteristic curve through P with the initial manifold is calculated as (Figure 4.24)

$$t_p = \frac{h - x_p}{v} \tag{4.32}$$

The solution of the Huxley–Zahalak equation is, moreover, required to be continuous. Taking all these facts into consideration (namely, straight characteristics, conditions specified at the initial manifold, continuity, and given probability distributions), the solution of the characteristic differential equation (4.26) is obtained by direct integration as:

$$n_p = \begin{cases} 0 & \text{for } t < \dfrac{h - x_p}{v} \\[1em] \dfrac{\alpha}{(\alpha + \beta)l_a}[1 - e^{((\alpha+\beta)/2v)(h^2 - (x_p + vt)^2)}] & \text{for } \dfrac{h - x_p}{v} \leqslant t < -\dfrac{x_p}{v} \\[1em] \dfrac{\alpha}{(\alpha + \beta)l_a}[1 - e^{((\alpha+\beta)/2v)h^2}]e^{(-g_0/v)(x_p + vt)} & \text{for } -\dfrac{x_p}{v} \leqslant t \end{cases} \tag{4.33}$$

These formulae, although obtained for one characteristic curve, actually furnish the complete solution $n(x, t)$. Indeed, given a point (x, t), we draw the unique characteristic through it. This line will intersect the vertical axis at $x_p = x - vt$. Substituting this value of x_p in the appropriate one of the three formulae (4.33), we obtain the value of $n(x, t)$. In this particular case, this procedure is tantamount to substituting x for $x_p + vt$ in the right-hand side of (4.33). We have used the general procedure to illustrate how the method of characteristics would work even if the specified contractile speed $v(t)$ were not constant in time. In the particular case when the speed is constant, we obtain the same result as Huxley [5], namely:

$$n(x, t) = \begin{cases} 0 & \text{for } x > h \\[1em] \dfrac{\alpha}{(\alpha + \beta)l_a}[1 - e^{((\alpha+\beta)/2v)(h^2 - x^2)}] & \text{for } 0 < x \leqslant h \\[1em] \dfrac{\alpha}{(\alpha + \beta)l_a}[1 - e^{((\alpha+\beta)/2v)h^2}]e^{-(g_0/v)x} & \text{for } x \leqslant 0 \end{cases} \tag{4.34}$$

Let us now use equations (4.34) and (4.29) to calculate the total force F. Note that, since t does not appear explicitly in any of the three formulae (4.34), $F(t)$ will turn out to be constant for a given constant speed of contraction, in agreement with the usual interpretation of Hill's experiments. The result of the integration required by equation (4.29) is obtained explicitly as

$$F = \frac{mAsk\alpha}{2(\alpha + \beta)l_a}\left[\frac{1}{2}h^2 + (1 - e^{((\alpha+\beta)/2v)h^2})\left(\frac{v}{\alpha + \beta} - \left(\frac{v}{g_0}\right)^2\right)\right] \tag{4.35}$$

Rheological and structural models

With a judicious choice of constants, this formula may be made to fit Hill's force–velocity law in the shortening range. In fact, by a similar treatment for energy and mechanical power, Huxley was able to match also some of Hill's thermodynamical considerations. Further developments along these lines were proposed by Huxley and Simmons [6] and by Zahalak and his associates [10, 11].

PROBLEMS

4.1 Show that the Maxwell body (Figure 4.25) abides by the differential equation:

$$\dot{u} = \frac{\dot{f}}{k} + \frac{f}{c} \quad (4.36)$$

4.2 Show that the Voigt body (Figure 4.26) abides by the differential equation:

$$f = ku + c\dot{u} \quad (4.37)$$

4.3 Show that the Kelvin body (Figure 4.27) abides by the differential equation:

$$k_s f + c\dot{f} = k_s k_p u + c(k_s + k_p)\dot{u} \quad (4.38)$$

4.4 How does each of the three models of Problems 4.1–4.3 respond to a sudden application of: (a) a constant force; and (b) a constant displacement?

4.5 Write a single differential equation governing the force–displacement behaviour of the model shown in Figure 4.28. What is the order of the highest derivative appearing in the equation?

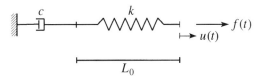

Figure 4.25 The Maxwell body.

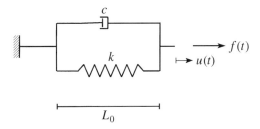

Figure 4.26 The Voigt body.

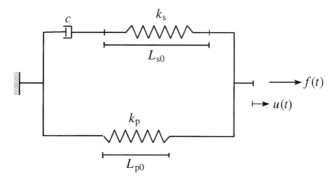

Figure 4.27 The Kelvin body.

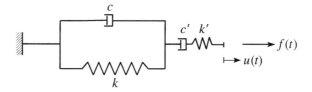

Figure 4.28 The four-element model.

4.6 What would the counterpart of equation (4.6) be if the prescription of equation (4.5), instead of (4.4), were used? What is the main difference between the two versions in terms of terminal velocities?

4.7 Write a computer code to implement equation (4.13). Verify the results plotted in Figures 4.9 and 4.11.

4.8 Modify the program of the preceding question by making the stiffness k_p of the parallel element depend on β and/or l_0. Experiment with different possibilities and comment on the effects on the overall force response.

4.9 Derive equations (4.15), (4.16), and (4.17).

4.10 Show that the left-hand side of equation (4.23), evaluated at a point (x, t), represents the directional derivative of $n(x, t)$ in the direction tangent to the characteristic curve (equation (4.24)) passing through that point.

4.11 Derive equations (4.33), (4.34), and (4.35).

4.12 Let the filaments slide past each other with constant acceleration α. What are the characteristic lines?

4.13 Why is it important that the probability of attachment grow away from the equilibrium position?

REFERENCES

[1] Allinger, T.L., Herzog, W. and Epstein, M. (1996) Stability of muscle fibres on the descending limb of the force-length relation: A theoretical consideration. *J. Biomech.*, **29**: 627–633.

[2] Courant, R. and Hilbert, D. (1966) *Methods of Mathematical Physics*, Vol II, Interscience, New York.
[3] Forcinito, M.A., Epstein, M. and Herzog, W. (1997) Don't give up on rheological models. Report no. 498, Department of Mechanical Engineering, University of Calgary.
[4] Hill, V.A. (1938) The heat of shortening and the dynamic constants of muscle. *Proc. R. Soc. London B*, **126**: 136–195.
[5] Huxley, A.F. (1957) Muscle structure and theories of contraction. *Proc. Biophys. Biophys. Chem.*, **7**: 255–318.
[6] Huxley, A.F. and Simmons, R.M. (1971), Proposed mechanism of force generation in striated muscle. *Nature*, **233**: 533–538.
[7] Mijailovich, S.M., Fredberg, J.J. and Butler, J.P. (1996) On the theory of muscle contraction: Filament extensibility and the development of isometric force and stiffness. *Biophys. J.*, **71**: 1475–1484.
[8] Pollack, G.H. (1990) *Muscles and Molecules: Uncovering the Principles of Biological Motion*, Ebner and Sons, Seattle.
[9] Zahalak, G.I. (1990) Modeling muscle mechanics (and energetics), in *Multiple Muscle Systems* (eds J.M. Winters and S.L.Y. Woo), Springer-Verlag, New York, pp. 1–23.
[10] Zahalak, G.I. and Ma, S.P. (1990) Muscle activation and contraction: Constitutive relations based directly on cross-bridge kinetics. *J. Biomech. Engng*, **112**: 52–62.
[11] Zahalak, G.I. and Motabar Zadeh, I. (1997) A re-examination of calcium activation in the Huxley cross-bridge model. *J. Biomech. Engng*, **119**: 20–29.

Part Two
Applications

5
Fundamentals of mechanics

5.1 INTRODUCTION

As suggested in Chapter 2, one of the possible levels at which skeletal muscle can be modelled consists of its representation as an assemblage of straight elements which remain straight upon deformation, and which elongate or contract in response to internally developed axial forces and/or neural activation. Because of the importance and relative simplicity of such models, we will devote this chapter entirely to a rigorous general formulation of their governing equations. As with any other deformable mechanical system, there are three aspects to be considered: a geometric description of the deformation, or **kinematics**; a statement of the physical laws at play, or **dynamics**; and a formulation of the behaviour of the materials involved, or **constitutive theory**. Before embarking on a detailed treatment of each of these three pillars of the theory, it might prove useful to discuss briefly their main features.

5.1.1 Kinematics

From the strictly kinematic point of view, the changing geometry of the system can be described in terms of a finite number of parameters called **degrees of freedom**. The free motion in space of a deformable straight segment, such as a muscle fibre, provided it remains straight at all times, is completely determined by six parameters which can be chosen as the three Cartesian coordinates of each of the two ends of the segment in some frame of reference (Figure 5.1). An assemblage of n such elements will, in principle, have a possible maximum of $6n$ degrees of freedom, but in practice this number is drastically reduced by the presence of **geometric constraints**. These constraints are of several types which we can, quite arbitrarily, group under three separate headings:

1. *Constraints of assemblage.* The different elements making up the system are interconnected at their end-points to conform to a topology that remains

116   *Theoretical models of skeletal muscle*

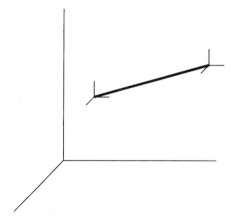

Figure 5.1 Element degrees of freedom.

fixed throughout the deformation process. Thus, for instance, a triangular arrangement of three bars will remain triangular, as shown in Figure 5.2, because the commonality of end-points will be preserved. These points of interconnectivity are usually called **nodes** or **joints**.

2. *External supports*. Some nodes will be externally supported, thus precluding the manifestation of some or all of the nodal displacement components. A typical example is shown in Figure 5.3.
3. *Other geometrical restrictions*. In muscle mechanics it is important to allow for other types of geometrical constraints to represent aspects not directly incorporated in the model. An important example is suggested by the intuitive notion that, because of the presence of water, the muscle is nearly incompressible. This intuitive notion has vast experimental support, as was shown in Chapter 2. In a two-dimensional analogue (Figure 5.4) incompressibility may be approximately translated as the preservation of the area of the panel. Notice that, in the example shown, that constraint is essential to guarantee the stability of the system, which might otherwise collapse into a triangle with the shorter fibre aligning itself with the upper membrane upon application of the slightest force f.

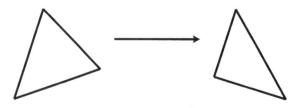

Figure 5.2 Topology preserved.

Fundamentals of mechanics

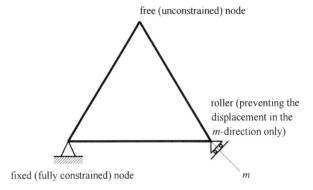

Figure 5.3 External supports.

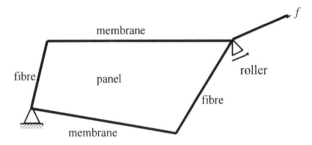

Figure 5.4 Preserving the panel area.

5.1.2 Dynamics

The dynamical laws governing the motion of the system are, basically, derivable from Newton's second law (force = mass × acceleration). But a few warnings are to be taken into consideration to avoid mistakes:

1. Newton's law as just stated applies to an idealized material particle, whereas here we have a system with distributed mass.
2. The internal forces arise, at least in part, from the deformation itself. The rigorous formulation of the laws of motion (or of the equilibrium equations, as the case may be), therefore, must involve angles and distances measured in the unknown deforming configuration. This gives rise to the so-called **geometrical nonlinearity**, which is essential in muscle mechanics (as opposed to many structural engineering applications) because the displacements are of the same order of magnitude as the gross muscular dimensions.
3. The presence of complicated geometrical constraints (such as incompressibility) may force the use of some of the tools of analytical mechanics (principle of virtual work, Lagrangian dynamics, etc.) because the forces necessary to maintain such constraints cannot easily be incorporated in a Newtonian

118 *Theoretical models of skeletal muscle*

framework. The importance of this last remark cannot be overemphasized, since attempts to obviate this route (by replacing it with a faulty intuition which endeavours to 'follow the distribution of forces', ignoring the less intuitive forces of constraint) may lead to serious errors in some proposed models.

In addition to the dynamical laws, a more sophisticated model may involve thermodynamics, chemistry, electromagnetism and control theory. These aspects are not included in the treatment of this chapter.

5.1.3 Constitutive theory

Whereas the two preceding points (kinematics and dynamics) involve, at most, mathematical difficulties, the proper representation of the material behaviour of a muscle fibre, or of a fibre bundle, or even of the simpler passive tendinous tissue, involves physical difficulties stemming, on the one hand, from the incompleteness of available experimental data and, on the other hand, from the unavailability of a consistent mathematical model that will accommodate all of the already generally accepted experimental results. This, then, is the weakest link of any model. The internal force developed in a muscle fibre seems to depend not only on the present values of the length, the speed of deformation and the activation, but also on the past history of those quantities in a manner that so far defies exact analysis. No model can be better that the worst of its components; but that model is best which incorporates exact kinematics and dynamics and which is flexible enough to allow for the implementation of the latest experimental findings. In other words, any computer implementation should be modular, so that a continual upgrading of the total model can be effected by plugging in different material models.

5.2 COORDINATES, DISPLACEMENT, AND ELONGATION

The measurement and analysis of elongations and other kinematic quantities is best accomplished by means of a **frame of reference**, also known as an **observer**.[1] One of the main properties of classical physical space is the existence of orthogonal frames, which can be used to generate global coordinate systems. Let such a frame be given once and for all. Then, the position of a point P relative to this frame is uniquely determined in terms of its three Cartesian coordinates X, Y and Z (Figure 5.5). Alternatively, one can identify the position of P by means of its **position vector**:

$$\mathbf{R} = \overrightarrow{OP} = X\mathbf{i} + Y\mathbf{j} + Z\mathbf{k} \tag{5.1}$$

---

[1] Physically meaningful quantities, however, must be (in a precise mathematical sense) independent of the observer. Thus, for example, a scalar variable (like the elongation of a bar) must have the same value for all observers, in spite of the fact that the coordinates measured by each observer are different.

Fundamentals of mechanics 119

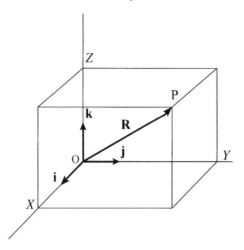

Figure 5.5 Reference frame.

Here O denotes the origin of the chosen frame, and i, j, k are unit vectors along the frame axes. By convention, we assume that the frame is right-handed, that is, if the extended right-hand thumb points upwards, like k, then the naturally bent fingers point from i to j along the shortest route (Figure 5.6). This rule is consistent with the vector cross-product equation:

$$k = i \times j \quad (5.2)$$

which for a left-handed frame would include a minus sign.

In a kinematic study of the motion of point P, one would like to distinguish

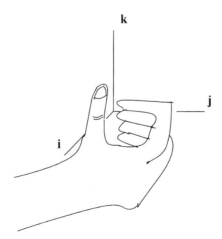

Figure 5.6 Right-handedness.

120   *Theoretical models of skeletal muscle*

between a **reference** (perhaps **initial**) position and one or more **spatial** (or **current**) positions of P. To facilitate the analysis, we will refer the spatial positions of P to the *same* frame as the referential position, but we will use lowercase letters to denote the point, its position vector and its spatial coordinates (Figure 5.7). Thus, the position vector of the current configuration p of P is expressed as:

$$\mathbf{r} = \overrightarrow{Op} = x\mathbf{i} + y\mathbf{j} + z\mathbf{k} \tag{5.3}$$

Since the current and initial position vectors are referred to the same frame, it makes sense to define their vector difference as:

$$\mathbf{u} = \mathbf{r} - \mathbf{R} = (x - X)\mathbf{i} + (y - Y)\mathbf{j} + (z - Z)\mathbf{k} = u_x\mathbf{i} + u_y\mathbf{j} + u_z\mathbf{k} \tag{5.4}$$

and to call it the **displacement vector** of point P. The displacement vector thus becomes our basic kinematic variable.

In straight-line models of skeletal muscle, tendons, aponeuroses, and fibres are represented by deformable straight bars whose lengths change with time in response to applied stimuli. Consider now a deformable straight bar (representing, say, a fibre) whose initial configuration is PQ and whose current configuration is pq (Figure 5.8). We would like to express the elongation (i.e. the change in length) *e* of the bar in terms of coordinates and displacements in as compact and elegant a manner as possible. One way to achieve this is to start by noticing that, by a direct application of Pythagoras's theorem, we can express the initial and current lengths, respectively, as:

$$L = [(X_Q - X_P)^2 + (Y_Q - Y_P)^2 + (Z_Q - Z_P)^2]^{1/2} \tag{5.5}$$

and

$$l = [(x_Q - x_P)^2 + (y_Q - y_P)^2 + (z_Q - z_P)^2]^{1/2} \tag{5.6}$$

In these formulae, the subscripts indicate the corresponding point, in an obvious

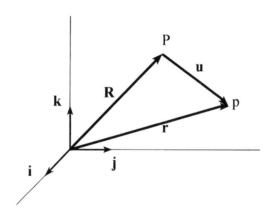

Figure 5.7 The displacement vector.

Fundamentals of mechanics

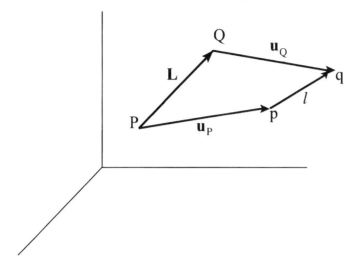

Figure 5.8 Kinematics of a fibre.

fashion. The elongation e is simply given by the difference

$$e = l - L \tag{5.7}$$

which, through equations (5.4), (5.5), and (5.6), turns out to be expressed in terms of the referential and current coordinates, or, equivalently, in terms of the referential coordinates and the displacement components of points P and Q.

The use of coordinate expressions tends to obscure the physical interpretation of the results. In the case of the elongation e, for example, a more illuminating way of expressing its magnitude is made possible by the introduction of the fibre-length vectors:

$$\mathbf{L} = \mathbf{R}_Q - \mathbf{R}_P \tag{5.8}$$

and

$$\mathbf{l} = \mathbf{r}_Q - \mathbf{r}_P \tag{5.9}$$

It follows from Figure 5.8

$$\mathbf{L} + \mathbf{u}_Q = \mathbf{u}_P + \mathbf{l} \tag{5.10}$$

whence

$$\mathbf{l} = \mathbf{L} + (\mathbf{u}_Q - \mathbf{u}_P) \tag{5.11}$$

Taking the dot product of each side of equation (5.11) with itself, we obtain

$$l^2 = L^2 + 2\mathbf{L} \cdot (\mathbf{u}_Q - \mathbf{u}_P) + (\mathbf{u}_Q - \mathbf{u}_P) \cdot (\mathbf{u}_Q - \mathbf{u}_P) \tag{5.12}$$

from which

$$e = [L^2 + 2\mathbf{L} \cdot (\mathbf{u}_Q - \mathbf{u}_P) + (\mathbf{u}_Q - \mathbf{u}_P) \cdot (\mathbf{u}_Q - \mathbf{u}_P)]^{1/2} - L \tag{5.13}$$

122 Theoretical models of skeletal muscle

This expression is completely equivalent to the previous one, contained in equations (5.4)–(5.7), but it reveals that the elongation is, albeit nonlinearly, the result of the relative displacement $\mathbf{u}_Q - \mathbf{u}_P$ of the ends of the bar. This relative displacement influences the elongation in two distinct ways: through its projection on the initial configuration of the fibre, and through the square of its own magnitude.

An important limiting case arises when the magnitude of the relative displacement is very small compared with the initial length of the bar.[2] In such a situation the (small) elongation e is to be obtained (by either of the preceding formulae) as the difference of two large numbers. This is computationally dangerous (although the high precision of today's computers makes it less so), and it is best to derive an approximate expression which yields e directly rather than as a difference. Taking into consideration the assumed smallness of the magnitude of $\mathbf{u}_Q - \mathbf{u}_P$, its square can be neglected. Moreover, recalling from calculus that the Taylor expansion yields the following expression for the square root of a number close to any given positive number a:

$$(a+h)^{1/2} = a^{1/2} + \tfrac{1}{2}h/a + o(h^2) \qquad (5.14)$$

where $o(h^2)$ stands for terms of order h^2 or smaller, we obtain the following linearized formula for the elongation:

$$e_L = (\mathbf{u}_Q - \mathbf{u}_P) \cdot \mathbf{L}/L \qquad (5.15)$$

Physically, this means that in the geometrically linearized ('small-displacement') theory, the elongation of a bar is simply given by the projection of the relative end-displacement on the original configuration of the bar. A more careful analysis would show that the linearized version gives correct answers even in the large-displacement régime, provided the rotations are small.

An interesting alternative formulation [1] that combines the elegance of the linearized version (for instance, no square roots and no difference between large numbers) with exactness, can be obtained by introducing a modified elongation variable e' defined as:

$$e' = e(1 + e/2L) \qquad (5.16)$$

Notice that for small elongations the contribution $e/2L$ will be negligible, so that e' reduces to the elongation[3] e. On the other hand, the following manipulations show that e' has a very elegant, rational, expression in terms of the relative

---

[2] Many problems involving the deformation of relatively stiff systems supported against gross rigid-body motions fall into this category. A good example is the metal frame used in muscle tests. The influence of the deformation of the frame can be estimated, if needed to correct measurements, using the small-displacement approximation.

[3] It is conceptually very important to distinguish between small displacements and small elongations (or strains). A case in point is the deformation of a bone. The bone may undergo very large displacements and rotations, but because of its relatively high stiffness, the strains will be very small. In this instance, e and e' become indistinguishable, but e_L gives completely wrong answers because it presupposes small rotations. (See Problems 5.1 and 5.2.)

Fundamentals of mechanics

end-displacement. Indeed, from equation (5.12), we may write:

$$(l+L)(l-L) = l^2 - L^2 = 2\mathbf{L} \cdot (\mathbf{u}_Q - \mathbf{u}_P) + (\mathbf{u}_Q - \mathbf{u}_P) \cdot (\mathbf{u}_Q - \mathbf{u}_P) \quad (5.17)$$

But, since $l - L = e$ and $l + L = (L + e) + L = 2L + e$, it follows that:

$$e(2L + e) = 2\mathbf{L} \cdot (\mathbf{u}_Q - \mathbf{u}_P) + (\mathbf{u}_Q - \mathbf{u}_P) \cdot (\mathbf{u}_Q - \mathbf{u}_P) \quad (5.18)$$

and, dividing through by $2L$, we obtain

$$e' = \mathbf{L} \cdot (\mathbf{u}_Q - \mathbf{u}_P)/L + \tfrac{1}{2}(\mathbf{u}_Q - \mathbf{u}_P) \cdot (\mathbf{u}_Q - \mathbf{u}_P)/L \quad (5.19)$$

which is the desired expression. It simply adds a quadratic term to the linearized version. On the other hand, it is an exact formula, valid for arbitrarily large displacements. Although these advantages make the use of the modified elongation potentially very attractive, there is no need to resort to it, since one can always use the exact expression as long as one is willing to deal with the square root. We have presented the two versions for completeness, and also for theoretical reasons that may later help the interested reader better to understand general continuum mechanical theories.[4]

5.3 RATES AND VIRTUAL DISPLACEMENTS

There are at least two reasons why we should be interested in infinitesimal changes in elongation brought about by infinitesimal displacements superimposed on a given spatial configuration. Indeed, it may happen that the material behaviour of the tendon or fibre includes a viscous component proportional to the instantaneous rate of elongation, which is obtained as the limit of the change in elongation divided by the time interval. But even in the purely elastic case, the infinitesimal variations themselves play an important role in the application of the principle of virtual work for deriving the equations of motion or of equilibrium of the system. For those two reasons, we will devote this short section to the derivation of the effect of small changes of end-point displacements on the elongation of a straight bar. We will denote such small displacements by $\delta\mathbf{u}$ and we will refer to them as **variations** or **virtual displacements**. One may also choose to regard them as the result of (fictitious or actual) velocities $\mathbf{v}$ acting over a small interval of time dt. Correspondingly, we will obtain expressions for the variations δe and δe_L of the true and the linearized elongations, respectively. (Alternatively, these expressions may be interpreted as being proportional to rates of elongation de/dt, etc.). Using equation (5.12) and differentiating equations (5.13) and (5.15), yields

$$\delta e = (\mathbf{L} + \mathbf{u}_Q - \mathbf{u}_P) \cdot (\delta\mathbf{u}_Q - \delta\mathbf{u}_P)/l \quad (5.20)$$

$$\delta e_L = \mathbf{L} \cdot (\delta\mathbf{u}_Q - \delta\mathbf{u}_P)/L \quad (5.21)$$

---

[4] The modified elongation variable turns out to be a one-dimensional version of the so-called right Cauchy–Green tensor of continuum mechanics in its full generality (see [2]).

124   *Theoretical models of skeletal muscle*

A similar calculation for the modified elongation yields

$$\delta e' = (\mathbf{L} + \mathbf{u}_Q - \mathbf{u}_P) \cdot (\delta \mathbf{u}_Q - \delta \mathbf{u}_P)/L \tag{5.22}$$

whereby we note that, while for calculating δe we divide by the present, unknown, length l (given by equation (5.12)), for calculating the modified counterpart $\delta e'$ we use the original length L. Both are legitimate, albeit different, measures of the variation of the elongation, and they obviously coincide in the limit of small strains, even in the large-deformation regime. The connection between equation (5.20) and (5.22) can also be obtained directly from equation (5.16) as

$$\delta e' = (1 + e/L)\delta e \tag{5.23}$$

5.4 EXTERNAL AND INTERNAL FORCES

Force being a primary concept in mechanics, its precise definition cannot be accommodated without recourse to an axiomatic approach. Nevertheless, even if we should, rather vaguely, agree to understand as a force any purely mechanical interaction, the distinction between two kinds of forces, external and internal, arises naturally. To understand this classification, let us define a **material system** as any definite collection of material particles, so that we may ascertain unambiguously whether any given particle in the universe belongs to the given system or to the rest of the universe, a third possibility being excluded. With this definition in hand, we are in a position to define an **external force** as any purely mechanical interaction that takes place between the system and the rest of the universe. **Internal forces**, on the other hand, are interactions between parts of the system itself, and are thus responsible, as it were, for keeping the system functioning as a unit. This distinction depends, of course, on what the system is in the first place. As an example, let us regard the Earth as the system under consideration (system I of Figure 5.9). Then the gravitational pull $\mathbf{F}_1$ of the Sun is certainly an external force. If, on the other hand, the Sun is considered as the system (system II of Figure 5.9), then the pull $\mathbf{F}_2$ of the Earth on the Sun is an external force to this system. Most natural forces abide by the so-called **principle**

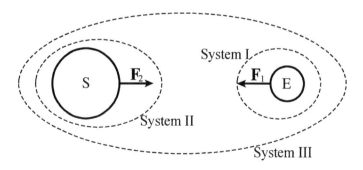

Figure 5.9 External and internal forces.

Fundamentals of mechanics

of action and reaction ('Newton's third law'), which states that given two separate systems, I and II say, the external force on II due to I is equal in magnitude and opposite in direction to the external force on I due to II. If, continuing with our example, we consider now the system consisting of both the Earth and the Sun (system III of Figure 5.9), the pair $\mathbf{F}_1, \mathbf{F}_2$ constitutes an internal force. Because the principle of action and reaction assures us that $\mathbf{F}_2 = -\mathbf{F}_1$ (as vectors), it follows that to specify the internal force representing the Sun–Earth interaction in system III it is sufficient to consider a single scalar quantity F, agreeing to call it positive when, say, it portrays an attraction, and negative otherwise.

A second example (Figure 5.10) is provided by a system consisting of two articulated bones actuated by one muscle. For the system as a whole, the external forces consist of $\mathbf{F}$, $\mathbf{R}$ and the couple C. The force developed within the muscle is an internal force. To make this internal force manifest, we mentally sever the muscle from the skeleton. In each of the two subsystems thus obtained (Figure 5.11), the muscle force N appears as an external force acting at each of the points where the cuts have been effected.

In muscle mechanics we deal with continuous systems, such as muscle fibres and tendons. Unlike the 'action-at-a-distance' internal forces of the planetary system, in a continuous system it is usually assumed that the internal forces are **localized** to the extent that: interactions between parts of the system take place only if those parts meet at a surface; and these interactions assume the form of forces of contact per unit area. This general idea gives rise to the concept of **stress**. For the particular case of a straight slender member acted upon by end forces only, it is enough to consider the total axial force acting on a cross-section assuming effectively that the stresses are uniformly distributed on the cross-section and are perpendicular to it. Figure 5.12a shows such a member attached at one end to a fixed support and pulled at the other end by an external force $\mathbf{F}$. To make internal forces manifest, we sever the system into two parts, L and R say, by means of an imaginary cross-sectional cut c–c (Figure 5.12b). For the

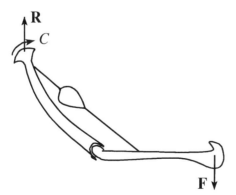

Figure 5.10 Bone–muscle system.

126   *Theoretical models of skeletal muscle*

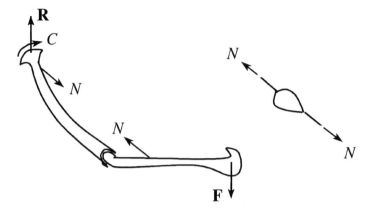

Figure 5.11 Bone and muscle subsystems.

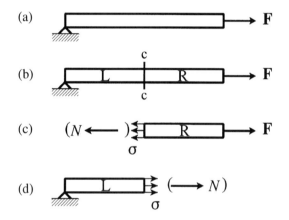

Figure 5.12 Forces in a continuous member.

part R, considered now as a new system, the effect of L becomes an external action exerted on the cross-section in the form of stresses σ per unit area (Figure 5.12c). The resultant of these stresses is an axial force N. A similar situation is true for part L, where the action of R is equal and opposite to the effect just discussed (Figure 5.12d). These forces permit the total original system L + R to remain 'glued' together at the cross-section c–c while transmitting the external force **F** to the support, much in the same way as the mutual gravitational pull permits the Sun–Earth system to remain functioning as such. In the case of the Solar system, we agreed to call the mutual interaction positive if it represented an attraction. Similarly, the internal axial force N is a scalar which will be considered positive if it represents a **tension**, such as in the examples of Figures 5.11 and 5.12, and negative if it corresponds to a **compression**.

In a case such as the bone subsystem of Figure 5.11, the bones can no longer

Fundamentals of mechanics

be considered as axially loaded, and a careful analysis of the stress distribution throughout the cross-sections leads to the introduction of other stress resultants in addition to the axial force N, such as bending moments, torsional moments, and shearing forces.

5.5 THE CONSTITUTIVE EQUATION

As we have seen in Chapter 4, muscle fibres respond to a history of elongation and activation by developing internal forces or stresses. As explained in the previous section, for the particular case of straight slender elements, it may be sufficient to represent the internal response in terms of a single scalar quantity N measuring the intensity of the internal axial force developed. We have agreed to consider this axial force positive if it represents a tension and negative if it represents a compression. In the tensile case, the internal forces react so as to prevent the two parts on either side of a cross-section from being pulled apart. A formula proposing a functional dependence of these internal forces on the history of the elongation and possibly of other variables is called a **constitutive equation**. Constitutive equations are ultimately subject to experimental verification. There are, however, certain general principles that they must satisfy, the most important being **observer indifference** and **thermodynamic consistency** (both of which are beyond the scope of this book). For the case of a muscle element, such as a fibre bundle, the constitutive law may look as complicated as

$$N(t) = f(e_{-\infty \to t}, \theta_{-\infty \to t}, g_{-\infty \to t}, a_{-\infty \to t}, \ldots) \quad (5.24)$$

where t is current time, and where $e_{-\infty \to t}, \theta_{-\infty \to t}, g_{-\infty \to t}, a_{-\infty \to t}, \ldots$ represent the whole past histories of the elongation, the temperature, the temperature gradient, the activation, ... More complicated constitutive equations can be conceived. A fibre is called **elastic** if its constitutive equation is of the simple form

$$N = f(e) \quad (5.25)$$

the variable t being omitted, it being understood that e is the present value of the elongation. An elastic fibre, then, has neither history dependence nor any sensitivity to temperature or other variables, except the present value of the elongation. This can be considered to be a good first approximation for some passive tissues such as tendon. A material for which the history dependence boils down to the additional appearance of the time derivative of the elongation is said to exhibit a **viscoelastic** behaviour. Thermal effects are included by means of the temperature alone (**thermal insulators**) and/or its gradient (**thermal conductors**). Appearance of a neural activation parameter signals active (as opposed to passive) behaviour, and so on. Needless to say, the formulation of suitable constitutive equations in muscle mechanics is still the greatest challenge for the researcher. For, whereas the mechanical principles governing equilibrium and motion are well understood and rigorously formulated, it has not yet been possible to find a single constitutive description which will adequately encompass

128 *Theoretical models of skeletal muscle*

the behaviour of muscle fibres under all known experimental conditions. Both the 'purely phenomenological' approach and the more fundamental 'structural' approach have so far failed to meet their goals. When using any constitutive model (such as Hill's or Huxley's) one should bear in mind their inherent limitations.

5.6 THE PRINCIPLE OF VIRTUAL WORK

When external forces or other external agents (e.g. neural activation) are applied to a deformable system such as a muscle, the different parts of the system undergo deformations which result, through the constitutive behaviour, in internal stresses. The interplay between external agents and internal forces is not arbitrary, since it is mediated and limited by two factors: the presence of geometrical constraints, such as supports or other constraints discussed in Section 5.1.1, which limit the degrees of freedom of the system; and the universal laws of mechanics, governing the motion (or equilibrium) of all conceivable material systems. Historically, the laws of mechanics were first formulated by Newton in the seventeenth century, a formulation which led to a spectacular success in its incredibly accurate description of planetary motion by means of a single formula ('force equals mass times acceleration') coupled with the law of universal gravitational attraction (the 'inverse square' law, also formulated by Newton). The application of Newton's picture of the universe to Earth-bound machines and other more mundane material systems met with equal success, and classical Newtonian mechanics is still at the very foundation of modern structural and mechanical engineering. During the eighteenth and nineteenth centuries alternative equivalent formulations of classical mechanics were proposed which came to be known under the general name of analytical mechanics. In the mechanics of deformable media the situation is similar, and there are well-developed formulations of both types. The tools of analytical mechanics are particularly well suited to systems with geometrical constraints. Because models of skeletal muscle are more often than not subject to rather involved geometrical constraints, we will present here a formulation of the **principle of virtual work** for deformable systems, one of the basic tools of the analytical approach, and we will show how with its help it is possible to obtain exact and energetically consistent equations of motion and/or equilibrium.

To understand the principle of virtual work, we must first introduce the notion of virtual displacement. A **virtual displacement** of a system is a set of small variations $\delta \mathbf{u}_i$ of its displacements, compatible with the geometrical constraints of the system. The precise mathematical meaning of 'smallness' is that the increment of any function of the displacements may be replaced by its differential. The fact that virtual displacements must not violate the geometrical constraints is essential to the further development of the theory, as we shall presently see. For definiteness, let the system under consideration consist of a number M of straight elements connected at a number m of nodes, some of which may be prevented

Fundamentals of mechanics

from movement by means of supports, and the remaining n may be subjected to external point-forces $\mathbf{F}_i$. We will assume, furthermore, that the members are very slender or that, if not slender, they are connected with each other at the common nodes by means of perfect hinges placed in such a way that each member can develop only an axial force N_I. Any given configuration of this system can be completely described by a set of actual displacements $\mathbf{u}_i (i = 1, ..., n)$ of any given magnitude and direction. As a result of these displacements, the elements of the system will undergo certain elongations $e_I (I = 1, ..., M)$ and sustain some internal forces N_I. Upon the imposition of a virtual displacement $\delta \mathbf{u}_i$, the members will undergo further virtual elongations δe_I, as per equation (5.20). The total **internal virtual work**, *IVW*, of the internal forces is then, by definition,

$$IVW = \sum_I N_I \, \delta e_I, \qquad I = 1, ..., M \qquad (5.26)$$

At the same time, the external forces $\mathbf{F}_i$ acting on the nodes will give rise to the **external virtual work** *EVW*, given by

$$EVW = \sum_i \mathbf{F}_i \cdot \delta \mathbf{u}_i, \qquad i = 1, ..., n \qquad (5.27)$$

Since the virtual displacements are, by definition, compatible with the geometrical constraints, the forces of reaction of the supports (or, for that matter, any of the forces necessary to maintain given geometrical constraints) automatically do not contribute to the external virtual work.

The principle of virtual work asserts that a system is in an equilibrium configuration if, and only if, the internal virtual work is identically equal to the external virtual work for all possible virtual displacements superimposed on the configuration:

$$IVW = EVW \qquad (5.29)$$

If one should include among the external forces also the 'forces of inertia' (i.e. the negative of the masses times the accelerations), then, according to **d'Alembert's principle**, the virtual work identity is equivalent to the equations of motion of the system. It is important to realize that the principle of virtual work prescribes that formula (5.28) as an identity rather than just an equation, so its validity extends over the whole range of possible virtual displacements. It is only by the force of this condition that a single scalar prescription may convey the same information as a whole system of equations, as we will see in some of the following examples.

Before embarking on the applications, some comments pertaining to the internal virtual work expression may be useful. The evaluation of the internal virtual work by means of equation (5.26) requires the use of equation (5.20), which contains a square root in the denominator involving the unknown displacement components. The use of equation (5.22) might in some cases be preferable, since it is a simple polynomial in the unknown displacements. By virtue of equation (5.22) we may write:

$$IVW = \sum_I N_I \, \delta e_I = \sum_I N'_I \, \delta e'_I, \qquad I = 1, ..., M \qquad (5.29)$$

where

$$N'_I = \frac{N_I}{1 + \dfrac{e_I}{L_I}} \qquad (5.30)$$

is a **modified axial force**,[5] the static dual of e'_I. Although this modified force does not have the immediate physical meaning of a 'physical' force pulling the bar internally, it is to be noted that: in the limit of small elongations, even at large displacements, it becomes equal to N_I; and, more importantly, any given constitutive equation in terms of N and e, can be transformed into an equivalent one in terms of N' and e', so that in practice the use of either pair of variables is justified. Finally, in the case of small displacements, the internal virtual work can often be evaluated using δe_L, a considerable simplification ('geometric linearization') which must be used with extreme caution.

Whether the physical, modified, or linearized expressions are used, through equations (5.20), (5.21), and (5.22), the internal virtual work will ultimately be obtained in terms of the virtual displacements themselves, so that by collecting terms it will eventually result in a formula of the type:

$$IVW = \sum_i [\text{coefficient}]_i \cdot \delta \mathbf{u}_i, \qquad (i = 1, ..., n) \qquad (5.31)$$

where $[\text{coefficient}]_i$ stands for the result of collecting all the terms affecting $\delta \mathbf{u}_i$. Equating this with the external virtual work equation (5.27), it follows that:

$$IVW = \sum_i ([\text{coefficient}]_i - \mathbf{F}_i) \cdot \delta \mathbf{u}_i \equiv 0, \qquad (i = 1, ..., n) \qquad (5.32)$$

identically for all virtual displacements $\delta \mathbf{u}_i (i = 1, ..., n)$ compatible with the constraints. If the virtual displacements correspond to actual degrees of freedom, then they can be given any values independently of each other. It follows, then, that their coefficients in equation (5.32) must independently vanish, thus yielding the equilibrium equations:

$$[\text{coefficient}]_i = \mathbf{F}_i, \qquad i = 1, ..., n \qquad (5.33)$$

If, on the other hand, there exist some geometrical constraints between the degrees of freedom, the preceding argument is no longer true and a modification of the method becomes necessary, as will be demonstrated in a later section. To

---

[5] This is roughly the one-dimensional version of the so-called second Piola–Kirchhoff stress tensor in general continuum mechanics.

Fundamentals of mechanics 131

illustrate the philosophy and the power of the method of virtual work, let us consider first the following idealized example.

5.7 EXAMPLE: A ONE-DEGREE-OF-FREEDOM SYSTEM

Two identical bars are assembled horizontally by means of hinges at A, B and C, as shown in Figure 5.13, and loaded with a vertical force P. The ends A and C are fully supported. Let us find the equilibrium equation and explore some particular cases.

This problem is geometrically nonlinear, since it is obvious that, in the initial geometry, it would be impossible for the (horizontal) axial forces in the bars to equilibrate a vertical force. The vertical deflection, therefore, will have to be large enough so as to generate a slope for the forces to be balanced. The constraints in this case are easily dealt with. Indeed, at points A and C all movement is precluded, and at point B, by symmetry, only the vertical component of the displacement will develop. These constraints result simply in the elimination of some displacements or displacement components and present, therefore, no difficulty. In later examples we will consider constraints consisting of rather involved relations between displacement components, and we will then present the means to handle them.

Denoting by v the vertical displacement of point B, we obtain, from equation (5.20), for bar AB:

$$\delta e = \frac{v\, \delta v}{(L^2 + v^2)^{1/2}} \tag{5.34}$$

where we have used $\mathbf{L} = L\mathbf{i}$, $\mathbf{u}_P = \mathbf{0}$, $\mathbf{u}_Q = v\mathbf{j}$. By symmetry, the total internal virtual work for the two bars is given by:

$$IVW = \frac{2Nv\delta v}{(L^2 + v^2)^{1/2}} \tag{5.35}$$

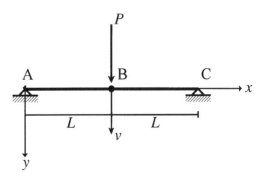

Figure 5.13 Idealized example.

with N representing the internal axial force in either bar. The external virtual work is:

$$EVW = P\,\delta v \tag{5.36}$$

According to the principle of virtual work, these expressions must be identically equal to each other for all δv, which immediately implies:

$$\frac{2Nv}{(L^2+v^2)^{1/2}} = P \tag{5.37}$$

which is the desired equilibrium equation. We note that, since, as can be seen from Figure 5.14,

$$\frac{v}{(L^2+v^2)^{1/2}} = \sin\alpha \tag{5.38}$$

the equilibrium equation, in this case, could have been derived by inspection. On the other hand, the method of virtual work delivers the right equations regardless of their complexity with equal elegance and exactness, as we shall see in other examples.

To solve for v we need to supplement equation (5.37) with a constitutive equation. If, for example, the material is elastic, the axial force N is a function $f(e)$ of the elongation e alone, which, according with equation (5.13), is given in this case by:

$$e = (L^2+v^2)^{1/2} - L \tag{5.39}$$

Introducing the elastic constitutive equation $N = f(e)$ in the equilibrium equation, we obtain:

$$\frac{2vf[(L^2+v^2)^{1/2} - L]}{(L^2+v^2)^{1/2}} = P \tag{5.40}$$

If, moreover, f happens to be a linear function, i.e.

$$N = ke \tag{5.41}$$

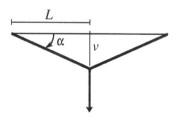

Figure 5.14 Deformed geometry.

Fundamentals of mechanics 133

with k a material constant, we obtain:

$$\frac{2vk[(L^2 + v^2)^{1/2} - L]}{(L^2 + v^2)^{1/2}} = P \tag{5.42}$$

which is an explicit nonlinear equation for v.

Had we used, instead of the physical variables N and e, the modified variables N' and e', we would have obtained from equations (5.22) and (5.29) the following (exact) expression for the internal virtual work:

$$IVW = 2N'v\delta v/L, \tag{5.43}$$

and the equilibrium equation would have become:

$$2N'v/L = P \tag{5.44}$$

which is completely equivalent to equation (5.37). An elastic constitutive equation

$$N' = g(e') \tag{5.45}$$

would have resulted in the final equation:

$$\frac{2vg(v^2/2L)}{L} = P \tag{5.46}$$

where equation (5.19) has been invoked. For a linear function

$$N' = ke' \tag{5.47}$$

the result is:

$$kv^3/L^2 = P \tag{5.48}$$

a simple cubic equation for v. This equation is as exact as equation (5.42), but it corresponds to a different constitutive behaviour. The reader can verify that if either constitutive equation is translated into its counterpart in the appropriate variables, the corresponding final equation is recovered. To appreciate how these two exact solutions of slightly different behaviours may differ from each other quantitatively, we present in Table 5.1 both solutions for different values of a non-dimensionalized load. For relatively small elongations the difference is insignificant.

Dynamic effects can be included in the model by adjoining to the external virtual work the virtual work of the forces of inertia. Thus, for example, if the mass of the system were to be lumped at the nodes, the addition to the *EVW* would be:

$$-m_B \frac{d^2v}{dt^2} \delta v \tag{5.49}$$

where m_B is the mass assigned to node B. Application of the principle of virtual

134   *Theoretical models of skeletal muscle*

Table 5.1 Comparative results for constitutive equations (5.41) and (5.47)

P/kL ($\times 10^{-3}$)	v/L (for $N = ke$)	v/L (for $N' = ke'$)	Difference (%)
0.9926	0.1	0.09975	0.25
7.7677	0.2	0.19805	0.98
25.3042	0.3	0.29358	2.14
57.2186	0.4	0.38534	3.66
105.5728	0.5	0.47263	5.47
171.0085	0.6	0.55506	7.49
253.0753	0.7	0.63253	9.64
350.6099	0.8	0.70514	11.86
462.0705	0.9	0.77310	14.10
585.7864	1.0	0.83672	16.33

work leads to the following equation of motion:

$$m_B \frac{d^2 v}{dt^2} + \frac{2Nv}{(L^2 + v^2)^{1/2}} = P \qquad (5.50)$$

a second-order ordinary differential equation for v as a function of the time t. Even without dynamics, the governing equation may become a differential equation (of the first order) if viscous effects are included in the constitutive behaviour via a function $N = f(e, de/dt)$. Finally, if the bars represented muscle fibres, which contain a contractile component regulated, say, by an activation parameter a, the constitutive equation might have the functional form $N = f(e, a)$. In this case, however, the equilibrium equation alone will not suffice to solve the problem unless the value of a is specified externally. Otherwise, a control model coupled with the mechanical one should be supplied.

5.8 A MORE GENERAL EXAMPLE

We wish to formulate the equilibrium equations for a plane assembly of straight hinged bars fully supported at some of the nodes. The intent is to come closer to the geometrical versatility needed for the representation of a muscle as an assembly of fibre bundles, aponeurotic membranes and tendons. We use the term 'bar' to refer to any such component of the assembly.

Let the assembly (Figure 5.15) consist of m nodes, n of which are free and the rest $(m - n)$ are fully supported. Without any loss of generality, we will number the nodes starting from the free ones, $i = 1, ..., n$. A virtual displacement, then, consists of a set of vectors $\{\delta \mathbf{u}_i\}$, $i = 1, ..., n$. Since the displacements $\mathbf{u}_i (i = 1, ..., n)$ are not geometrically constrained, the virtual displacement

Fundamentals of mechanics 135

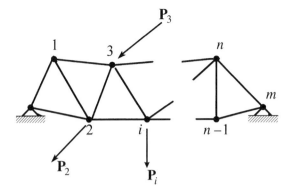

Figure 5.15 A planar bar assembly.

vectors $\delta\mathbf{u}_i (i = 1, \ldots, n)$ can be given any arbitrary values, independently of each other. On the other hand, the displacements $\mathbf{u}_j (j = n + 1, \ldots, m)$, being prescribed as zero by the supports, do not have a virtual displacement counterpart.

Let e_{ij} denote the elongation of the segment $i - j$, and N_{ij} the axial force in that segment. If there is no bar joining i with j, we will consider N_{ij} to be identically zero. Then, the internal virtual work over a virtual displacement set $\{\delta\mathbf{u}_i\}$, $i = 1, \ldots, n$, is given by:

$$IVW = \sum_i \sum_j N_{ij}\, \delta e_{ij}, \qquad i = 1,\ldots,n, \qquad j = i+1,\ldots,m \qquad (5.51)$$

with δe_{ij} given by equation (5.21) with the obvious change of notation, i.e.

$$IVW = \sum_i \sum_j \frac{N_{ij}(\mathbf{L}_{ij} + \mathbf{u}_j - \mathbf{u}_i)\cdot(\delta\mathbf{u}_j - \delta\mathbf{u}_i)}{l_{ij}}, \qquad i = 1,\ldots,n, \quad j = i+1,\ldots,m$$

(5.52)

The external virtual work is:

$$EVW = \sum_i \mathbf{P}_i \cdot \delta\mathbf{u}_i, \qquad i = 1,\ldots,n \qquad (5.53)$$

where, for simplicity, we assume that the nodal forces $\mathbf{P}_i$ do not change during the deformation process. According to the principle of virtual work, the difference between the last two expressions is identically zero at equilibrium for all values of the virtual displacements compatible with the supports. Since the support conditions are already taken into consideration by the exclusion of the higher $m - n$ nodes, these virtual displacements can independently attain any values, and it follows that the coefficients of each $\delta\mathbf{u}_i$ must independently vanish. For, suppose that the coefficient of, say, $\delta\mathbf{u}_1$ (in the difference

136   *Theoretical models of skeletal muscle*

$IVW - EVW$) is not zero. Then, we may choose $\delta \mathbf{u}_1 \neq \mathbf{0}$ and $\delta \mathbf{u}_2 = \delta \mathbf{u}_3 = \cdots = \delta \mathbf{u}_n = \mathbf{0}$, leading to a contradiction. Notice that the essence of this argument is the independence of the virtual displacements. If there existed a constraint between them, the argument would no longer apply, since the special choice made above might not be permissible.

To complete this example, we need only identify the coefficient of a generic $\delta \mathbf{u}_k (k = 1, .., n)$ in the virtual work expressions above. In the EVW expression, the coefficient is obviously $\mathbf{P}_k$. As for the IVW expression, we obtain, by first fixing $\delta \mathbf{u}_i$ and then $\delta \mathbf{u}_j$:

Coefficient of $\delta \mathbf{u}_k$ in IVW

$$= \sum_j \frac{N_{kj}(\mathbf{L}_{kj} + \mathbf{u}_j - \mathbf{u}_k)}{l_{kj}} + \sum_i \frac{N_{ik}(\mathbf{L}_{ik} + \mathbf{u}_k - \mathbf{u}_i)}{l_{ik}}, \quad j = k+1, \ldots, m, \ i = 1, \ldots, k-1$$

$$= \sum_i \frac{N_{ik}(\mathbf{L}_{ik} + \mathbf{u}_k - \mathbf{u}_i)}{l_{ik}}, \quad i = 1, \ldots, m, \ i \neq k \quad (5.54)$$

where we have used $\mathbf{L}_{ij} = -\mathbf{L}_{ji}$, and $N_{ij} = N_{ji}$. The equilibrium equations are

$$\sum_i \frac{N_{ik}(\mathbf{L}_{ik} + \mathbf{u}_k - \mathbf{u}_i)}{l_{ik}} = \mathbf{P}_k, \quad i = 1, \ldots, m, \ i \neq k \quad (5.55)$$

for each $k = 1, \ldots, n$. This is a set of $2n$ scalar equations for the $2n$ unknown components of the n free displacements $\mathbf{u}_i (i = 1, \ldots, n)$. The method of virtual work has once again shown its power in producing error-free equilibrium equations in a manner free of any intuitive considerations. Geometrical and physical intuition may now be used to reaffirm our certainty on the validity of the equations obtained: the left-hand side of equation (5.55) provides simply the sum of all the internal forces converging at node k, each one multiplied by the corresponding unit vector in the direction of the corresponding deformed bar.

In closing this example, we note that the equilibrium equations (5.55), in spite of their neat appearance, are nonlinear in the displacements for two reasons: the constitutive equation introduces a further dependence on the displacements; and the denominators l_{ik} are actually functions of the displacements, as per equation (5.12). The latter nonlinearity can be alleviated by using the modified variable formulation. Equations in terms of the modified variables can be obtained in a similar way by means of the principle of virtual work.

5.9 GEOMETRIC CONSTRAINTS

We have, so far, encountered only constraints of support, which result generally in the exclusion of one or more of the displacement components from the list of degrees of freedom. In practice, however, constraints (such as the incompressibility of skeletal muscle) may appear which establish an interdependence

Fundamentals of mechanics

between several degrees of freedom, that is, a functional relation of the form

$$\phi(u_1, ..., u_n) = 0 \tag{5.56}$$

to be satisfied identically. In some cases, it may be possible to read off from this equation one of the displacement components in terms of the others and to replace every appearance of this component with the expression thus obtained, thereby eliminating it from the problem. If that can be accomplished, then the problem is effectively reduced to an unconstrained one with one less degree of freedom. More often than not, however, this elimination is either impossible or inconvenient. In such cases, one of the ways to tackle the problem is by means of the so-called **method of Lagrange multipliers**. It consists of adding a new variable ('Lagrange multiplier'), λ, to the list of degrees of freedom and then replacing the principle of virtual work by the following principle:

$$IVW - EVW \equiv \delta(\lambda\phi) \tag{5.57}$$

which expresses mathematically the principle of virtual work in the presence of a geometric constraint – a deformable system in the presence of a geometric constraint (5.56) is in equilibrium if, and only if, the difference between the internal and the external virtual work is identically equal to the variation of the product $\lambda\phi$ for all possible independent variations $\delta u_1, ..., \delta u_n, \delta\lambda$.

If more constraints are imposed, the prescription is similar. Let p denote the number of constraints:

$$\phi_\alpha(u_1, ..., u_n) = 0, \quad \alpha = 1, ..., p \tag{5.58}$$

We then introduce p Lagrange multipliers, $\lambda_\alpha (\alpha = 1, ..., p)$, and stipulate:

$$IVW - EVW \equiv \sum_\alpha \delta(\lambda_\alpha \phi_\alpha), \quad \alpha = 1, ..., p \tag{5.59}$$

The Lagrange multipliers turn out to have the following physical meaning: they are proportional to the 'forces' necessary to maintain the constraints. But we notice that the nature of these 'forces' is not always intuitively clear.

On developing equation (5.57) in the spirit of equation (5.32), we obtain:

$$\sum_i ([\text{coefficient}]_i - \mathbf{F}_i) \cdot \delta\mathbf{u}_i \equiv \lambda \sum_i \left(\frac{\partial \phi}{\partial \mathbf{u}_i}\right) \cdot \delta\mathbf{u}_i + \phi\delta\lambda, \quad i = 1, ..., n \tag{5.60}$$

whence it follows that

$$[\text{coefficient}]_i = \mathbf{F}_i + \lambda \frac{\partial \phi}{\partial \mathbf{u}_i}, \quad i = 1, ..., n \tag{5.61}$$

and

$$\phi = 0 \tag{5.62}$$

138 Theoretical models of skeletal muscle

Equations (5.61) and (5.62) constitute a system of $3n + 1$ simultaneous scalar equations for the $3n + 1$ scalar unknown quantities $\mathbf{u}_1, \ldots, \mathbf{u}_n, \lambda$. For the general case of p constraints, we obtain:

$$[\text{coefficient}]_i = \mathbf{F}_i + \sum_\alpha \lambda_\alpha \frac{\partial \phi_\alpha}{\partial u_i}, \qquad \alpha = 1,\ldots,p, \qquad i = 1,\ldots,n \quad (5.63)$$

and

$$\phi_\alpha = 0, \qquad \alpha = 1,\ldots,p \quad (5.64)$$

a system of $3n + p$ equations for the $3n + p$ scalar unknowns $\mathbf{u}_1, \ldots, \mathbf{u}_n, \lambda_1, \ldots, \lambda_p$.

Although the method of Lagrange multipliers results in an increase in the number of equations, it has the advantage (over a more direct technique of elimination, say) of incorporating the unaltered form of the unconstrained equations, thus avoiding errors in the formulation and permitting the addition, removal and/or modification of constraints with relatively little effort.

5.10 EXAMPLE: PRESERVATION OF AREA

We wish to formulate and solve the equilibrium equations for the idealized unipennate muscle model shown in Figure 5.16, assuming that the area of the panel remains unchanged throughout the deformation process. For the sake of analytical simplicity, in this example we consider only passive elastic behaviour and we adopt the geometrically linearized theory. These unrealistic assumptions for this particular example will yield, exceptionally, an explicit analytic solution, thus allowing for a full discussion of the physical meaning of the constraint equation and its repercussions in terms of the dangers of predicting internal forces intuitively.

This particular constraint has been briefly discussed and physically motivated in Section 5.1. Here we attempt an explicit solution. The suggestion of using the geometrically linearized ('small-displacement') formulation in no way implies a limitation of the method, but is adopted mainly for pedagogical reasons, since it allows us to obtain explicit results without the use of a computer. The geometrical symmetry of the structure has been adopted only for that same reason.

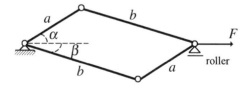

Figure 5.16 Unipennate example data.

Fundamentals of mechanics

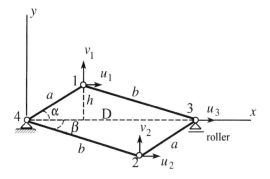

Figure 5.17 Nodal displacements.

The first step of the solution consists of numbering the nodes and the corresponding degrees of freedom. This is shown in Figure 5.17. Next, we express the constraint mathematically in terms of those degrees of freedom. The original distance D between points 4 and 3 is given by

$$D = a \cos \alpha + b \cos \beta \qquad (5.65)$$

The original height h of point 1 above the x-axis is likewise given by

$$h = a \sin \alpha = b \sin \beta \qquad (5.66)$$

which, by the way, shows that of the four geometrical data a, b, α, β, only three need to be specified. The deformed counterparts of D and h are, respectively, $D + u_3$ and $h + v_1$. Similarly, the vertical distance of point 2 below the x-axis becomes $h - v_2$. The initial and final areas are, therefore,

$$A_{\text{initial}} = Dh \qquad (5.67)$$

and

$$A_{\text{final}} = \tfrac{1}{2}(D + u_3)(h + v_1) + \tfrac{1}{2}(D + u_3)(h - v_2) \qquad (5.68)$$

The constraint $A_{\text{final}} - A_{\text{final}} = 0$, then, becomes

$$\phi = \tfrac{1}{2}(D + u_3)(2h + v_1 - v_2) - Dh = 0 \qquad (5.69)$$

or, simplifying and neglecting higher order terms,

$$\phi = \tfrac{1}{2}D(v_1 - v_2) + hu_3 = 0 \qquad (5.70)$$

The term neglected is $\tfrac{1}{2}u_3(v_1 - v_2)$, which is small compared with the terms retained. This order-of-magnitude consideration is entirely consistent with the assumptions already made for the geometrically linearized theory. In an exact, geometrically nonlinear, formulation, the neglected term would have to be retained. Continuing in the spirit of the linearized theory, we use equation (5.15)

140 Theoretical models of skeletal muscle

to calculate the elongations, thereby obtaining:

$$e_{14} = u_1 \cos \alpha + v_1 \sin \alpha \tag{5.71}$$

$$e_{13} = -u_1 \cos \beta + v_1 \sin \beta + u_3 \cos \beta \tag{5.72}$$

$$e_{23} = -u_2 \cos \alpha - v_2 \sin \alpha + u_3 \cos \alpha \tag{5.73}$$

$$e_{24} = u_2 \cos \beta - v_2 \sin \beta \tag{5.74}$$

and the variations

$$\delta e_{14} = (\delta u_1) \cos \alpha + (\delta v_1) \sin \alpha \tag{5.75}$$

$$\delta e_{13} = -(\delta u_1) \cos \beta + (\delta v_1) \sin \beta + (\delta u_3) \cos \beta \tag{5.76}$$

$$\delta e_{23} = -(\delta u_2) \cos \alpha - (\delta v_2) \sin \alpha + (\delta u_3) \cos \alpha \tag{5.77}$$

$$\delta e_{24} = (\delta u_2) \cos \beta - (\delta v_2) \sin \beta \tag{5.78}$$

The internal virtual work is given by:

$$\begin{aligned} IVW &= N_{14} \delta e_{14} + N_{13} \delta e_{13} + N_{23} \delta e_{23} + N_{24} \delta e_{24} \\ &= [N_{14} \cos \alpha - N_{13} \cos \beta] \delta u_1 + [N_{14} \sin \alpha + N_{13} \sin \beta] \delta v_1 \\ &\quad + [-N_{23} \cos \alpha + N_{24} \cos \beta] \delta u_2 + [-N_{23} \sin \alpha - N_{24} \sin \beta] \delta v_2 \\ &\quad + [N_{13} \cos \beta + N_{23} \cos \alpha] \delta u_3 \end{aligned} \tag{5.79}$$

and the external virtual work is, clearly,

$$EVW = F \delta u_3 \tag{5.80}$$

According to the principle of virtual work in the presence of a constraint, we must have:

$$IVW - EVW \equiv \delta(\lambda \phi) \tag{5.81}$$

identically for all independent variations δu_1, δv_1, δu_2, δv_2, δu_3, $\delta \lambda$. Using our particular constraint, equation (5.70), we obtain:

$$\delta(\lambda \phi) = [\tfrac{1}{2} D(v_1 - v_2) + h u_3] \delta \lambda + [\tfrac{1}{2} \lambda D] \delta v_1 + [-\tfrac{1}{2} \lambda D] \delta v_2 + [\lambda h] \delta u_3 \tag{5.82}$$

Equating the coefficients of the corresponding variations, we end up with the following system of six simultaneous equations for the six unknown quantities u_1, v_1, u_2, v_2, u_3 and λ:

$$N_{14} \cos \alpha - N_{13} \cos \beta = 0 \tag{5.83}$$

$$N_{14} \sin \alpha + N_{13} \sin \beta = \tfrac{1}{2} \lambda D \tag{5.84}$$

$$-N_{23} \cos \alpha + N_{24} \cos \beta = 0 \tag{5.85}$$

$$-N_{23} \sin \alpha - N_{24} \sin \beta = -\tfrac{1}{2} \lambda D \tag{5.86}$$

$$N_{13} \cos \beta + N_{23} \cos \alpha = \lambda h + F \tag{5.87}$$

$$\tfrac{1}{2} D(v_1 - v_2) + h u_3 = 0 \tag{5.88}$$

Fundamentals of mechanics

Naturally, in order to solve this system we need to specify a constitutive equation. To make matters as simple as possible in this illustrative example, let us assume linear elasticity, i.e.

$$N_{ij} = k_{ij} e_{ij} \tag{5.89}$$

where k_{ij} are constants representing, albeit imperfectly, the (passive) stiffnesses of the fibres and aponeuroses. If we assume, furthermore, that

$$k_{14} = k_{23} = K_1 \tag{5.90}$$

and

$$k_{13} = k_{24} = K_2 \tag{5.91}$$

it is not difficult to foresee that the solution will be symmetric, with

$$v_2 = -v_1 \tag{5.92}$$

and

$$u_2 = u_3 - u_1 \tag{5.93}$$

Equation (5.83) becomes then equivalent to (5.85), and equation (5.84) is the same as (5.86). The system of equations to be solved is, therefore:

$$K_1 e_{14} \cos \alpha - K_2 e_{13} \cos \beta = 0 \tag{5.94}$$

$$K_1 e_{14} \sin \alpha + K_2 e_{13} \sin \beta = \tfrac{1}{2}\lambda D \tag{5.95}$$

$$K_1 e_{14} \cos \alpha + K_2 e_{13} \cos \beta = \lambda h + F \tag{5.96}$$

$$D v_1 + h u_3 = 0 \tag{5.97}$$

Instead of substituting, as we should, the expressions (5.71) and (5.72) for e_{ij} in terms of u_1, v_1 and u_3, we see that, in this case, the first three equations can be solved for e_{13}, e_{14} and λ, yielding

$$e_{14} = \tfrac{1}{2}(F/K_1)\cos \beta /[\cos \alpha \cos \beta - (h/D)\sin(\alpha + \beta)] \tag{5.98}$$

$$e_{13} = \tfrac{1}{2}(F/K_2)\cos \alpha /[\cos \alpha \cos \beta - (h/D)\sin(\alpha + \beta)] \tag{5.99}$$

$$\lambda = (F/D)\sin(\alpha + \beta)/[\cos \alpha \cos \beta - (h/D)\sin(\alpha + \beta)] \tag{5.100}$$

By virtue of equation (5.66) we have

$$[\cos \alpha \cos \beta - (h/D)\sin(\alpha + \beta)] = \cos(\alpha + \beta) \tag{5.101}$$

so that we may write

$$e_{14} = \tfrac{1}{2}(F/K_1)\cos \beta /\cos(\alpha + \beta) \tag{5.102}$$

$$e_{13} = \tfrac{1}{2}(F/K_2)\cos \alpha /\cos(\alpha + \beta) \tag{5.103}$$

$$\lambda = (F/D)\tan(\alpha + \beta) \tag{5.104}$$

Finally, using these values in conjunction with equations (5.71), (5.72), and (5.88), we can obtain the displacement components. In particular, the resulting

142   *Theoretical models of skeletal muscle*

formula for u_3 is

$$u_3 = \frac{\frac{1}{2}F(\cos^2\beta/K_1 + \cos^2\alpha/K_2)}{\cos^2(\alpha+\beta)} \qquad (5.105)$$

which shows that the equivalent stiffness of the elastic system at point 3 is:

$$K_{eq} = \frac{2\cos^2(\alpha+\beta)}{\cos^2\beta/K_1 + \cos^2\alpha/K_2} \qquad (5.106)$$

a formula which could hardly have been obtained intuitively.

So as to gain some intuitive insight into the meaning of the solution, we must imagine how this constrained situation might be physically attained. We could conceive of a (two-dimensional) nearly incompressible 'fluid' which fills the space between the bars. Moreover, the physical bars must have a very high flexural stiffness so that they can withstand the pressure developed in the fluid while remaining straight. In the limit, as the fluid becomes truly incompressible and the bars flexurally stiff, we should obtain our idealized solution. The Lagrange multiplier, then, is a measure of the fluid pressure P. If we wish to check that this is the case by means of a free-body equilibrium diagram, we must be very careful in the choice of the free body, since upon cutting the bars at an arbitrary point we will in general encounter bending moments and shear forces, which have been excluded *ab initio* from the idealized hinged model. This example shows the dangers of relying on intuition alone for the solution of problems involving geometrical constraints. On the other hand, intuition is an excellent guide when assessing the validity of a solution obtained through dry mathematical routine. In this example we could reason as follows: by symmetry, the shear forces at the mid-point cross-sections of the bars are zero. Therefore, a legitimate free-body diagram is the one shown in Figure 5.18, where M_1 and M_2 are bending moments. We enforce in this free-body diagram the vanishing of the sum of forces in the vertical direction, namely:

$$-N_{14}\sin\alpha - N_{13}\sin\beta + \tfrac{1}{2}PD = 0 \qquad (5.107)$$

from which we obtain for the fluid pressure P the value:

$$P = (N_{14}\sin\alpha + N_{13}\sin\beta)(2/D) \qquad (5.108)$$

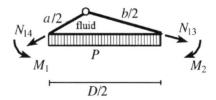

Figure 5.18 Free-body diagram.

Fundamentals of mechanics

which, comparing with equation (5.84) shows that:

$$\lambda = P \tag{5.109}$$

In other words, the physical meaning of the Lagrange multiplier is that of a quantity proportional to the 'force' necessary to maintain the constraint, in this case the internal pressure of the 'fluid'. But notice that the constraint itself, equation (5.70), could have been multiplied by any arbitrary non-zero constant without altering its geometrical meaning, so that, correspondingly, the Lagrange multiplier would have had to be divided by that same constant to keep the virtual work contribution unchanged, resulting effectively in an indeterminacy of the magnitude of the Lagrange multiplier.

Recall, finally, the main reason for having presented this example, namely: to show that a seemingly straightforward reasoning based on the direct decomposition of forces yields a wrong result because it ignores the subtle contribution of the constraint forces. Indeed, if the free-body diagram of node 3 (with the vertical reaction obviously vanishing) were taken (incorrectly) to be the one shown in Figure 5.19, a direct equilibrium reasoning would yield the following values for the internal forces:

$$N'_{23} = \frac{F \sin \beta}{\sin(\alpha + \beta)}$$
$$N'_{13} = \frac{F \sin \alpha}{\sin(\alpha + \beta)} \tag{5.110}$$

The correct values, according to equations (5.102) and (5.103), are:

$$N_{23} = \frac{F \cos \beta}{2 \cos(\alpha + \beta)}$$
$$N_{13} = \frac{F \cos \alpha}{2 \cos(\alpha + \beta)} \tag{5.111}$$

Notice that the correct values show that, surprisingly perhaps, equilibrium cannot be achieved if $\alpha + \beta = 90°$, i.e. when the fibres are perpendicular to the aponeuroses (and the panel area is preserved). Moreover, if $\alpha + \beta > 90°$, then both the fibres and the aponeuroses would have to be in compression while the

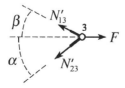

Figure 5.19 Free-body diagram of node 3, ignoring contributions of forces of constraint.

144 Theoretical models of skeletal muscle

Table 5.2 Data for Problem 5.2

Time	u_{Px}	u_{Py}	u_{Pz}	u_{Qx}	u_{Qy}	u_{Qz}
0.1	0.734	1.146	0.397	−3.710	1.608	0.321
0.2	0.946	1.089	0.584	−7.069	2.352	0.834
0.3	2.578	−3.457	0.928	−9.987	2.576	1.218
0.4	7.485	−6.786	1.238	−11.764	5.326	1.457

'fluid' would be in tension! These results are, of course, subordinate to having assumed infinitesimal displacements (geometrically linearized theory). Nevertheless, they give sufficient indication of how unreliable a reasoning based upon misguided intuition can be.

PROBLEMS

5.1 Show that for a large rigid-body rotation of a bar (for instance, 90° in a plane), the exact elongation, as well as the modified elongation, vanish, whereas the linearized elongation does not.

5.2 In an experiment aimed at quantifying fibre shortening during an *in vivo* motion of a cat muscle–tendon complex, the displacements of the end-points P and Q of a fibre of the cat medial gastrocnemius muscle were measured very accurately by means of high-speed video, yielding the results (in millimetres) shown in Table 5.2. The initial coordinates of the end-points P and Q are, respectively, $x_P = 19.7$ mm, $y_P = -3.5$ mm, $z_P = 0.0$ mm, and $x_Q = 41.0$ mm, $y_Q = 5.1$ mm, $z_Q = 0.0$ mm. For each of the times listed, calculate (i) the exact elongation e; (ii) the modified elongation e'; and (iii) the linearized elongation, all with respect to the original configuration at the initial time. Comment on the range of applicability of the linearized theory, with particular attention to the magnitude of the rotations.

5.3 Derive the equilibrium equations for the system of Figure 5.13 assuming that the force P is no longer vertical, but forms an angle of 60° with bar BC.

5.4 For the system of Figure 5.13, assume that the force P is vertical, but node B is forced to move on an incline forming 15° with the vertical direction. Solve this problem directly and by using a Lagrange multiplier, and show that the two solutions are identical. What is the physical meaning of the Lagrange multiplier?

REFERENCES

[1] Epstein, M. and Tene, Y. (1971) Nonlinear analysis of pin-jointed space trusses. *Am. Soc. Civ. Engrs J. Struct. Div.*, **97**: 2189–2202.
[2] Truesdell, C. and Toupin, R.A. (1960) *The classical field theories*, in: *Handbuch der Physik* (ed. S. Flügge), Vol. III/1, Springer-Verlag, Berlin, pp. 226–793.

6
Towards a complete muscle model

6.1 INTRODUCTION

Having presented the basic theoretical tools for the correct formulation of the mechanical aspects of a straight-line representation of skeletal muscle, we now proceed to show how these tools may be put into practice to produce a viable model for the numerical simulation of real-life problems. Rather than advocating any particular model, our aim is to demonstrate how any proposed model can be implemented in a manner consistent with the general principles of mechanics and of numerical analysis. This approach will be exhibited by the explicit development of a complete functioning model, down to its smallest details.

Even the simplest realistic model will necessitate the use of a computer. In Appendices A and B, we will discuss general methods for tackling linear and nonlinear equations and for integrating time-dependent problems. Our computer codes will be developed fully only for planar deformations, with the understanding that spatial models constitute a relatively straightforward extension. Indeed, in going from two to three dimensions, nothing essential changes in the methodology of either the formulation of the equations or their numerical solution. Moreover, the representation of the deformation of a muscle by that of its medial plane (which is, roughly, an approximate plane of symmetry containing the tendons) is a reasonable approximation in many cases. Perhaps the strongest compromise in a planar model is the way in which volume conservation is enforced. For, whereas in a spatial model the volume of any desired portion of the muscle can be calculated exactly at any state of deformation, in a planar model one has to make do with an area or with some other *ad hoc* two-dimensional measure. Since in the previous chapter we have thoroughly discussed the Lagrange multiplier technique for handling any geometric constraint, we have deemed it sufficient to implement just one type of constraint in our computer code, leaving other constraints as an exercise for the reader. Our choice of constraint (panel area preservation) is complicated enough as a demonstration and has the added advantage of being the most direct two-dimensional analogue of true volume preservation.

6.2 A PROGRAM FOR STATIC ANALYSIS OF SKELETAL MUSCLE

The general architecture of a computer program can be best understood in terms of three main components acting in succession (Figure 6.1) and commonly known as the **pre-processor, main processor** and **post-processor**. The pre-processor accepts the geometrical and loading data and prepares them in a manner suited to the analysis to be carried out by the main processor. The results are then cast in tabular and graphical fashion by the post-processor. For obvious reasons, commercial packages emphasize the pre- and post-processing phases, and the 'user friendliness' of a program is usually judged on the basis of their sophistication. But in reality the esssence of a program lies in the main processing phase, where the analysis and numerical approximation procedures take place. We will, therefore, focus our attention on the description of a complete, albeit less than perfect, main processor of a program for skeletal muscle modelling, leaving improvements of the input/output aspects of the code to be better developed by experienced programmers. In Appendix A, complete listings of all the routines are given (in QBASIC) and the entire program can be run on any personal computer. Readers are encouraged to use these routines as building blocks which may serve as a starting point for their own improvements and modifications. Notice that Appendix A contains also all the relevant theory pertaining to each of the program routines. A detailed reading of this appendix is recommended to those who wish to gain a clearer understanding of the passage from the principle of virtual work to the actual calculations necessary to obtain a numerical solution. In muscle mechanics this passage is rather difficult due to the many geometric and material nonlinearities involved.

In addition to the coding of the equilibrium equations and the constraint equations, an essential element for the completion of a fully fledged program for the analysis of a muscle model is a routine containing the constitutive law. For illustration purposes, we will adopt the following quadratic force–length relation giving the normal stress (in newtons per square millimetre) in a fully activated

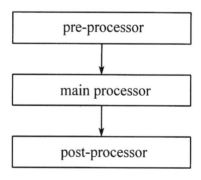

Figure 6.1 General program architecture.

fibre as a function of the ratio

$$r = \frac{L}{L_0} \tag{6.1}$$

between the fibre length L and the length L_0:

$$f(r) = -0.772r^2 + 1.544r - 0.494 \tag{6.2}$$

The range of physical validity of this law is $0.4 < r < 1.6$. The actual force in an isometrically activated fibre is given by

$$N = Aaf(r) \tag{6.3}$$

where A is the fibre cross-sectional area and a the activation parameter ($0 \leqslant a \leqslant 1$). In addition to this active force, a passive component (which develops gradually for lengths above optimal), could be easily incorporated into this force–length relation. More important, though, is the realization that *the force–length relation (even at full activation) does not represent the behaviour of muscle fibres as the muscle is stretched*. Indeed, formula (6.2) represents a curve-fitting exercise over a number of different experiments, rather than a force response throughout a single gradual deformation process. In particular, the softening behaviour that would be implied by the presence of the descending limb ($r > 1$) of the curve, with its inherent instability, is not observed experimentally. On the contrary: upon further small stretching or shortening at full activation, the force–length response always follows a line of positive slope. The exact representation of this behaviour, as well as the response of partially activated fibres, is still a matter of research. For the purpose of illustration, we adopt here the following constitutive equation which captures some of the main features of the response:

$$N = Aa[f(r_0) + k(r_0)(r - r_0)] \tag{6.4}$$

where r_0 is the length of the fibre upon *first* activation, and $k(r_0)$ is a length-dependent positive slope for active elongations beyond that length. For illustration purposes we choose $k(r_0) = $ constant $= 0.9264$, which is the value of the initial slope of the ascending limb. We stress that we are not advocating the use of this particular constitutive equation, but only suggesting that, since it captures some of the essential features of the experimentally observed behaviour, it is a good starting point for numerical computations. It should be replaced by an improved equation based on available and new experimental data.

As far as the tendinous tissue is concerned, we will adopt the following quadratic constitutive law [1] in the tensile range:

$$\sigma = 15\,160[(\varepsilon + 7.25 \times 10^{-4})^2 - 5.26 \times 10^{-7}] \tag{6.5}$$

where σ is the stress (N/mm^2) and ε is the strain, that is, the elongation divided by the length of the original force-free configuration. The computer code

148 Theoretical models of skeletal muscle

implements these constitutive laws in the FUNCTION CONS. The interactive input makes the program virtually self-explanatory. The output is written onto the file STATIC.OUT. When running this program, one should keep in mind the inherent nonlinearities involved. Not every set of data will lead to convergence, so that an intelligent monitoring of the physical meaning of the sequence of loading is indispensable for success.

6.3 EXAMPLE: STATIC DEFORMATION OF A CAT MEDIAL GASTROCNEMIUS MUSCLE

The initial geometry of a cat medial gastrocnemius is given in Figure 6.2. Tables 6.1 and 6.2 show the measured mid-plane coordinates of the nine points marked in the figure and the measured tendon forces and corresponding estimated global

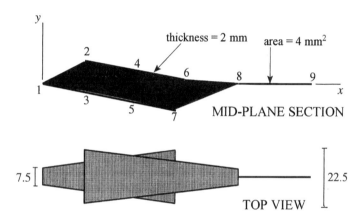

Figure 6.2 Initial gastrocnemius geometry.

Table 6.1 Mid-plane coordinates of the points marked in Figure 6.2

Point	x (mm)	y (mm)
1	0.0	0.0
2	18.1	8.5
3	19.7	−3.5
4	41.0	5.1
5	39.4	−6.9
6	64.1	1.1
7	59.0	−10.4
8	88.0	0.0
9	120.0	0.0

Towards a complete muscle model 149

Table 6.2 Measured tendon forces and global activation coefficients

Load (N)	Activation
10	0.1
20	0.2
50	0.6
100	1.0

activation coefficients. We wish to determine the deformed muscle geometry for each of the loads and corresponding activation levels.

We will divide the body of the muscle into three panels (Figure 6.3), using the given measured points as dividers. (It is also recommended to subdivide into more panels, by interpolation, in order to check the reliability of the results.) Since each of the four model fibres (1–2, 3–4, 5–6, 7–8) will be representative of a whole bundle of fibres, we will lump the corresponding volumes by assigning to each model fibre an effective cross-section, whose area can be estimated by dividing the corresponding volume of influence by the average fibre length. We assume each volume of influence to extend half-way along the distance to the neighbouring fibres. Since the total volume of the muscle is approximately 11 250 mm^3, and the average fibre length is 25 mm, the effective cross-sectional area for each of the two inner fibres can be taken as 150 mm^2, with half that amount alloted to each of the two outer fibres. Appendix A contains a verbatim reproduction of an interactive session with the program as well as a listing of the resulting output file.

A glance at the numerical results shown in Appendix A reveals that the displacements have turned out to be small everywhere. The reason for this behaviour can be traced to the fact that the effects of contraction due to the increasing levels of activation have been in each case almost completely neutralized by the elongations produced by the constantly increasing force. To see that this has been the case, we have considered a second loading example in which the external force is kept fixed at 10 N while the activation increases as before. The numerical results for this case are also listed in Appendix A. Figure 6.4 shows

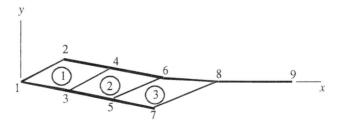

Figure 6.3 Subdivision into panels.

150   *Theoretical models of skeletal muscle*

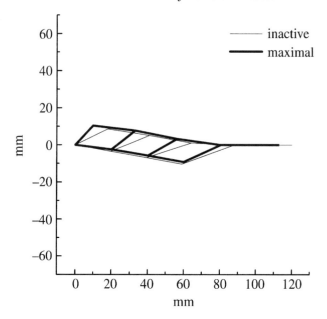

Figure 6.4 Muscle deformation.

the final geometry, corresponding to maximal activation, compared with the initial geometry of the muscle. Dramatic changes in angles of pinnation (up to about 21°) and in fibre lengths (up to about −28%) are observed.

6.4 TIME-DEPENDENT MODELLING

Time dependence in muscle mechanics may arise from two different sources: velocity dependence of the fibre response; and inclusion of inertia effects. It is generally considered that inertia effects are of secondary importance, either because the masses involved at the fibre level are relatively small or because the (poorly known) internal damping makes their effects very short-lived. Nevertheless, when including the unquestionably important velocity dependence of the fibre response, we will keep at the same time the effects of inertia, thus achieving a unified presentation. We note that keeping the effects of inertia leads to second-order differential equations, thereby permitting the independent specification of initial lengths and velocities, which would otherwise be related to each other by the (first-order) equilibrium equations at the initial time.

A word of caution is called for before embarking on the numerical solution of time evolution problems. There is an essential difference in the nature of the equations that govern the *statics* of a system with a finite number of degrees of freedom as opposed to those that govern its *dynamics*. In the case of statics, the equations are *algebraic*. This means that no matter how complicated these

Towards a complete muscle model

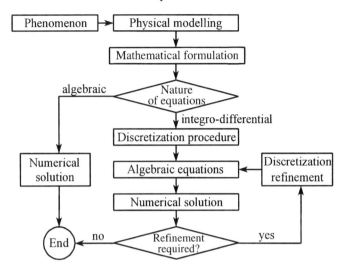

Figure 6.5 Approximate solutions.

equations are, once an approximate solution is obtained such that the residuals are very small, we can be reasonably sure that we are very close to an exact solution of the problem. In the dynamic case, on the other hand, the equations governing the evolution of the system are *differential*, that is, they are equations involving (time) derivatives. Solutions of such equations are not just numbers, but functions (of time). Any approximative procedure involves, therefore, two steps: one in which the differential equations themselves are replaced by algebraic equations, and the other in which these algebraic equations are solved. The first step is usually referred to as a **discretization** procedure, and its necessity stems from the fact that, with very few exceptions, the exact solutions of systems of differential equations cannot be expressed in terms of known functions. Any discretization procedure involves a potential degradation of the quality of the solution, for, even if the resulting algebraic equations were to be solved exactly, one would still have only obtained an exact solution to the discretized version of the equations, and the problem would remain as to how to assess the validity of such a solution as an approximation to the original differential equations. That is why one normally provides, along with the discretization scheme, a **convergence criterion** involving some sort of increasing refinement of the discretization procedure. These concepts are summarized in Figure 6.5.

6.5 A PROGRAM FOR TIME-DEPENDENT ANALYSIS OF SKELETAL MUSCLE

The force–velocity dependence can be incorporated approximately by introducing a multiplicative Hill's-law-like factor into the constitutive law (6.4). For

152 Theoretical models of skeletal muscle

illustrative purposes, and to avoid complications due to lack of differentiability or zero stiffness beyond the maximal speed of contraction, in the present version of the program we will approximate the force–velocity dependence with the following exponential formula:

$$F(v) = 1 - \tanh\left(\frac{v}{120}\right) = 1 - \frac{(e^{v/120} - e^{-v/120})}{(e^{v/120} + e^{-v/120})} \qquad (6.6)$$

where v is speed (positive if contractile) in millimetres per second. We stress once more that, with little extra effort, any other constitutive law can be substituted for this one by implementing the corresponding changes in the function CONS.

The numerical treatment of the time dependence is effected by means of **Houbolt's method**, described in Appendix B. The inertia effects can be ignored, if so wished, by setting the masses to zero. A complete listing of the computer code is given in Appendix B, where the relevant theory is discussed in detail. The results are stored in the file DYNAMIC.OUT.

6.6 EXAMPLE: TIME-DEPENDENT DEFORMATION OF A CAT MEDIAL GASTROCNEMIUS MUSCLE

Using exactly the same data as for the static example, we assume now that the loading steps are spaced at regular 50 ms intervals. After the first 200 ms, the total load is sustained for another 200 ms. We exclude inertia effects by setting all masses to zero. A complete reproduction of the interactive session and the corresponding output file are given in Appendix B. Comparing the results with their static counterparts, we observe that it has taken an extra 200 ms of sustained load to attain the static results (cf. the dynamic results at step 8 with the corresponding static results at step 4). The tendon tip has crept an extra 0.6 mm during that time period before attaining the static deformation of 4.3 mm.

REFERENCES

[1] Prilutsky, B.I., Herzog, W., Leonard, T.R. and Allinger, T.L. (1996) Role of the muscle belly and tendon of soleus, gastrocnemius, and plantaris in mechanical energy absorption and generation during cat locomotion. *J. Biomech.*, **29**: 417–434.

7
Movement control

7.1 INTRODUCTION

Skeletal muscle is arguably the most important component of the musculo-skeletal system from a biomechanics point of view. Skeletal muscles produce force, they are the primary moment producers around joints, and, as such, they produce the voluntary movements of animals and humans. Skeletal muscles are also largely responsible for the loading of joints and bones in static and dynamic tasks. However, if one's interest in skeletal muscle stopped here, then this chapter would not be necessary. Here, we deal with the aspect of movement control, or more specifically the (force–time) interaction of synergistic muscles in the production of voluntary movements. Basic aspects of the neurophysiology and anatomy of movement control are discussed, followed by theoretical and experimental considerations on movement control.

This chapter will be restricted to considerations on movement control at a single joint. It is realized that this is a severe restriction, and certain results obtained from single-joint experiments may not be generalizable to the control of entire limb or body movements. This restriction was necessary, however, because there are no experimental data of simultaneous, individual force recordings from muscles across more than a single joint. Since there is no comprehensive and generally accepted paradigm of movement control, and therefore, most proposed mechanisms of movement control are largely speculative, we wanted to stay on firm ground in this discussion, at least in terms of the experimental evidence on movement control that is presently available.

7.2 THE NEUROPHYSIOLOGY OF MOVEMENT CONTROL

Movements occur at joints, and joints are typically crossed by muscles or their corresponding tendons. Anatomically, joints are places where bones meet. Aside from the muscles, joints typically contain a capsule and ligaments which, together with the shape of the articulating surfaces, determine to a large degree the type and extent of movement that is possible at the joint. Although the forces

154   Theoretical models of skeletal muscle

in ligaments, joint capsules, and contact regions may produce a moment about a joint, and therefore contribute to joint movement, muscles are the primary producers of movement, as well as being the only active producers of movement and being under direct voluntary control.

Force (and moment) control in a skeletal muscle is not only governed by input from the brain but also influenced by control units on the spinal level (so-called pattern generators) and input from afferent (sensory) pathways originating from cutaneous receptors, joint receptors, and proprioceptors in skeletal muscles: the muscle spindles and Golgi tendon organs (Figure 7.1). The input from all these sources reaches the α motor neurones on the spinal level which activate the muscles: the larger the number of activated α motor neurones and the higher the

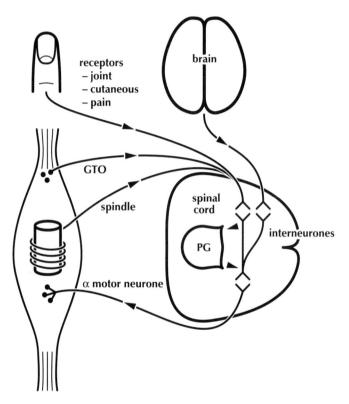

Figure 7.1 Schematic representation of neural pathways which influence muscle force production. The α motor neurone transmits the nerve signal for muscular contraction. The larger the number of active α motor neurones and the higher their corresponding firing frequency the bigger the muscular force. α motor neurone firing is influenced by descending pathways from the brain, motor pattern generators (PG) on the spinal level, and afferent pathways originating from a variety of feedback systems – for example, cutaneous pathways, pain receptors, joint receptors, muscle spindles and Golgi tendon organs (GTO).

firing frequency of the activated motor neurones, the larger the muscle force for given contractile conditions.

The synaptic input to the α motor neurones, be it from descending pathways of the motor cortex in the brain, pattern generators, or afferent neurons, is typically not direct but is channelled through interneurones (Figure 7.1). These interneurones act as switches with variable thresholds; they turn signals on or off. The summation of all these inputs forms, in some unknown way, the signal transmitted along the α motor neurones which innervate the muscle and produce contraction.

Alpha motor neurones are not the only efferent pathway for skeletal muscle control; there is also the γ motor neurone system. The γ system innervates muscle spindles which are proprioceptors sensitive to stretch and rate of stretch. Muscle spindles consist of nerve endings that are wrapped around modified muscle fibres (intrafusal fibres) (Figure 7.2) and they are arranged in parallel to the regular (extrafusal) muscle fibres. Typically, several intrafusal fibres are enclosed in a connective tissue capsule that makes up the muscle spindle. The end parts of the muscle spindle can contract, and it is these contractions that are controlled by γ motor neurones. Since muscle spindles are sensitive to stretch and the rate of stretch, muscle spindles may be activated in two ways: either by stretching the entire muscle, or by contraction of the end parts of the muscle spindle through γ motor neurone activation which stretches the central part of the spindle, and so produces a perceived stretch of the proprioceptive part of the spindle.

In theory, a movement can be initiated by activation of the α motor neurones which would produce direct contraction of the muscle, or by activation of the γ motor neurones which would cause contraction of the spindles, produce a spindle

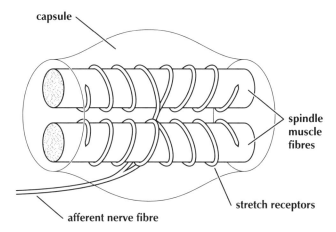

Figure 7.2 Schematic structure of a muscle spindle. In the centre of the spindle are stretch receptors wrapped around the intrafusal (muscle spindle) fibres. The ends of the intrafusal fibres (not shown) can contract.

156   *Theoretical models of skeletal muscle*

stretch, and cause activation of the α motor neurones triggered by the spindle afferent response. Initiating or controlling movements via the α–γ motor neurone system probably requires changes in the roles of the two systems depending on the task the muscle is asked to perform. In principle, a muscle can shorten or lengthen, and movements can occur actively or passively (Figure 7.3a). It has been proposed, and there is experimental evidence, that the α–γ interactions are task-specific [19], and the tasks may be divided into four groups: active shortening, passive shortening, active lengthening, and passive lengthening.

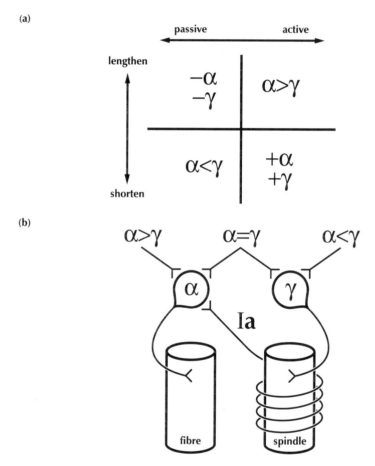

Figure 7.3 (a) Schematic illustration of the active/passive and shortening/lengthening functions of a muscle and the corresponding α and γ motor neurone interactions. A minus sign indicates absence of α or γ activity; a plus sign indicates activity. α > γ indicates strong activity in the α and little or no activity in the γ pathways, and vice versa for α < γ. (b) Schematic illustration of the α and γ systems and their association with muscle fibres and muscle spindles. (Adapted from [19].)

For active shortening of a muscle, it is proposed that there is coactivation of the α–γ system so that spindle firing is preserved during shortening and any perturbation (stretch) can be detected readily by the spindles (Figure 7.3b). For passive shortening, α motor neurones are not activated but the γ system needs to be selectively recruited to maintain spindle activity at non-zero levels (i.e. prevent spindles from becoming slack). For active lengthening (eccentric contractions), α motor neurones are selectively recruited (depending on the force demands), whereas γ motor neurones are not activated, otherwise the spindle output would probably be saturated and movement control would be impaired. Finally, for passive lengthening, the α–γ system is not activated. In this situation, the spindles presumably are sensitive detectors of the passive stretch of the muscle.

Summarizing, it should be realized that α motor neurone activation, which is responsible for muscle stimulation, depends on a variety of inputs: descending pathways from the brain, spinal pattern generators, afferent signals from muscle proprioceptors (the muscle spindles and Golgi tendon organs), as well as further afferent signals from a variety of receptors located in the joint capsule or skin and free nerve endings found in ligaments. Furthermore, there are two efferent pathways which can produce muscle activation, the α and the γ motor neurones. The α motor neurones directly activate the muscle fibres, the γ motor neurones activate the muscle spindles and cause shortening of the spindles at the ends, thereby stretching the stretch-sensitive central part. This stretch causes muscle spindle firing and causes α motor neurone activation. The exact and detailed pathway of muscle force control is not well understood, and a comprehensive paradigm of force control in skeletal muscle that is applicable across muscles and tasks is presently not available.

7.3 THE ANATOMY OF MOVEMENT CONTROL

In the mechanical analysis of movement, neurophysiological control models have rarely been used. The complexity of assigning a neurophysiologically-based system of control equations to multiple muscles contributing to joint or limb movement has rendered these approaches difficult for practical applications. Engineers and biomechanists have typically approached the problem of movement control by stating that an intersegmental resultant joint moment must be satisfied by the moments of the individual muscle forces produced about the joint:

$$\mathbf{M} = \sum_{i=1}^{n} (\mathbf{r}_i \times \mathbf{F}_i) \tag{7.1}$$

The resultant moment, $\mathbf{M}$, and the moment arm vectors, $\mathbf{r}$, can typically be determined using high-speed video analysis and anatomical information, respectively. Therefore, the only unknowns in equation (7.1) are the muscle forces, or, more

158 Theoretical models of skeletal muscle

precisely, the magnitudes of the muscle forces since the lines of action of the muscle forces can normally be determined from the anatomy of the system of interest.

If there was exactly one muscle for each degree of rotational freedom of the joint, then equation (7.1) would be a determined mathematical system and the time history of muscle force production could be calculated uniquely. However, most human and animal joints are crossed by more muscles than there are rotational degrees of freedom in the joint, thus producing an underdetermined set of equations, i.e. a set of equations which contains more unknowns (muscle forces) than system equations.

Consider, for example, flexion–extension at the human elbow. Flexion–extension corresponds to one rotational degree of freedom, however, there are (at least) three primary elbow flexors (the biceps, brachialis, and brachioradialis) and one primary elbow extensor (the combined heads of the triceps) (Figure 7.4). Rewriting equation (7.1) in scalar form for this particular case gives

$$M_E = \sum_{i=1}^{4} d_i \cdot F_i \qquad (7.2)$$

where M_E is the resultant flexor–extensor moment about the elbow, and d_i and F_i are the ith moment arm (common perpendicular of the elbow joint axis and the line of action of the ith muscle) and muscle force magnitude, respectively. Assuming again that M_E can be calculated and the moment arms can be determined based on anatomical considerations, equation (7.2) represents one equation with four unknowns; the four muscle force magnitudes. Knowing these four muscle force magnitudes at any given instant in time and for any movement requires a complete knowledge of the control of the elbow joint. Equation (7.2) is a mathematically redundant system that has an infinite number of possible solutions.

Although the system shown above is mathematically redundant, there has been much debate whether the system is also functionally redundant. Obviously,

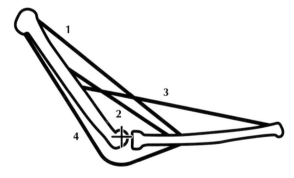

Figure 7.4 Schematic illustration of a human elbow with the three primary elbow flexors and the one primary elbow extensor.

Movement control 159

removing one of the muscles from the system would restrict the tasks that could be performed. Removing the single elbow extensor (Figure 7.4) would prevent elbow extension against resistance; removing one of the flexors would reduce maximal elbow flexor strength and possibly limit maximal elbow flexion speed. However, many normal, everyday elbow flexion movements could still be performed if one of the three primary flexors was gone.

As a second example, consider the human knee. The human knee has one primary rotational degree of freedom, flexion–extension, however, internal–external rotation and abduction–adduction movements are also possible within certain limits. Considering the knee as a system of three rotational degrees of freedom, the moment equations for the knee allow for the unique determination of three muscle force magnitudes. However, the knee is crossed by many more than three muscles. The primary knee extensors include four muscles (the three vasti and the rectus femoris), and the primary knee flexors include four muscles (the semitendinosus, semimembranosus, and the two heads of the biceps

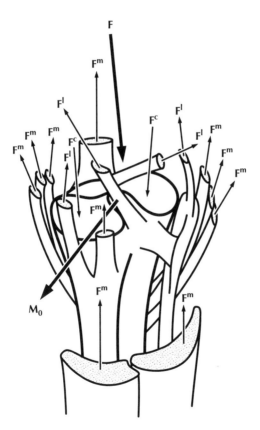

Figure 7.5 Schematic illustration of the potential force-carrying structures across the knee. Adapted from [5].)

femoris). Other muscles crossing the knee include the popliteus, gastrocnemius, plantaris, gracilis, and sartorius.

The muscles crossing the human knee primarily cause flexion–extension. Internal–external rotation and abduction–adduction at the knee are largely restricted by capsular and ligamentous forces since the congruency of the articular surfaces of the knee (tibiofemoral joint) does not provide much stability. Considering all these potential force–carrying structures, the resultant, intersegmental moment about the knee, $\mathbf{M}_k$, might be written as:

$$\mathbf{M}_k = \sum_{i=1}^{m} (\mathbf{r}_i^m \times \mathbf{f}_i^m) + \sum_{j=1}^{l} (\mathbf{r}_j^l \times \mathbf{f}_j^l) + \sum_{k=1}^{c} (\mathbf{r}_k^c \times \mathbf{f}_k^c) + \sum_{p=1}^{o} (\mathbf{r}_p^o \times \mathbf{f}_p^o) \quad (7.3)$$

where $\mathbf{r}$ and $\mathbf{f}$ designate location vectors and forces of internal structures, respectively, the superscripts m, l, c, and o refer to muscles, ligaments, (bony) contacts, and other, respectively, and the summation counters m, l, c, and o stand for the number of muscles, ligaments, (bony) contacts, and other force carrying structures crossing the knee. Even presuming that the instantaneous location vectors and lines of action of all force vectors can be determined based on anatomical considerations, calculating the unknown force magnitudes would still be a formidable task, one that has not been solved satisfactorily to date. The anatomical complexity of the knee is illustrated in Figure 7.5. Although not all force-carrying structures are shown in this particular figure, it helps to perceive that there are in excess of 20 major force carrying structures which cross the knee and should be considered in a complete mechanical and control model of this particular joint.

7.4 THEORETICAL AND EXPERIMENTAL CONSIDERATIONS ON MOVEMENT CONTROL

Much work in movement control has concentrated on identifying the role of isolated control pathways in reduced animal preparations. For example, the cat ankle extensor muscles have been used extensively to elucidate the details of spinal motor pattern generators and to study the influence of muscle spindle pathways on muscle force and stiffness. Although such studies provide tremendous insight into isolated pathways of movement control, the interaction of these pathways with other control systems in a fully intact preparation is not well understood.

There are very few 'mechanical' approaches to movement control. Movement control from a mechanical perspective may be defined as the precise interaction of the force–time histories of all muscles contributing to a movement or controlling the moment at a given joint. For example, if the muscle forces crossing the knee (Figure 7.5) were known at any instant in time during a movement (e.g. walking), then one could argue to have a complete description of the active (moment) control of that particular joint. There is no model that has gained general acceptance in the scientific community for predicting individual muscle forces accurately across a joint. Furthermore, to our knowledge, there is no joint (animal or human) for which all muscle forces controlling the joint have been measured simultaneously (except perhaps, if one accepts estimates of force magnitudes from joints controlled by a single muscle, such as the muscle controlling shell closure in clams).

A theoretical solution to calculating the individual muscle forces contributing to joint control is associated with solving equations of the form of equations (7.1)–(7.3). As stated before, such systems of equations are typically underdetermined (or redundant) and they typically have an infinite number of possible solutions. In order to illustrate the problems associated with a redundant mathematical system, imagine one equation which contains two unknowns, for example $x \cdot y = 12$. Obviously, this equation has many valid solutions (e.g. $x = 1$, $y = 12$; $x = 2$, $y = 6$; $x = 120$, $y = 0.1$); however, the question of importance in movement control at a joint is whether there is one unique solution (i.e. a set of muscle forces) for a given situation (i.e. a given movement), or whether a given movement is performed with completely different sets of muscle forces. The experimental consensus on this question is that well-trained, repetitive movements (such as locomotion) are performed in a stereotypical way. For example, force measurements performed in cat ankle extensor muscles during locomotion showed amazingly consistent patterns across animals as observed in different laboratories [1, 12, 14, 24]. Therefore, it can be assumed safely, at least to a first approximation, that well-learned, cyclic movements are performed with a precisely specified and repeatable set of individual muscle force–time histories. It has been demonstrated that the small, step-by-step variations of the ankle extensor forces observed in cat locomotion can be associated with systematic step-by-step variations in kinematic parameters of the corresponding hindlimb [13].

Once it has been accepted that a given movement is always performed using a stereotyped pattern of muscle force–time histories, the question arises, how this specific pattern is chosen from all the possible patterns. In order to answer this question, one has to ensure first that the mathematical redundancy translates into a functional redundancy. Is the human or the animal musculoskeletal system, or at least the control of joint moments, really redundant from a functional perspective, i.e. can a certain movement and a certain joint moment really be achieved with largely varying muscle forces? A casual glance at equations (7.1)–(7.3) would suggest such redundancy, but equations (7.1)–(7.3) do not contain functional properties of the muscles; they just contain the fact that muscles can produce force.

In the following, I would like to illustrate two examples based on experimental observations in the cat ankle extensor muscles which might raise (at least some) doubt in the reader's mind as to whether the functional system (cat ankle movement and moment control) is really as redundant as equations (7.1)–(7.3) might suggest. Let us consider first the so-called paw shake response. Paw shaking in the cat can be elicited by fixing a piece of adhesive tape to the paw of the hindlimb; the cat tries to rid itself of the tape by shaking its paw vigorously at a frequency of about 8–10 Hz. Ankle extension in this situation could be accomplished by activating all ankle extensors; however, it has been found that the soleus is not activated for paw shaking, while the gastrocnemius is [1, 23]. There are a variety of reasons why the soleus may not be activated in this situation; the one most often proposed is that the soleus is composed primarily of slow-twitch fibred motor units, therefore it is unable to produce adequate force during the limited time of ankle extension during the paw shake response. Although this particular argument has not been proven beyond a doubt, it illustrates that the soleus and gastrocnemius are very different muscles in terms of fibre type distribution, and that there might be movements which can easily be performed by one of the muscles but not by the other. Although not proven, it is fair to assume that the cat could not produce a normal paw shake response using the soleus only. Therefore, from a functional point of view, ankle extension during a paw shake response might not be a functionally redundant movement, although from a mathematical point of view, performing ankle extension with the soleus only is a perfectly acceptable solution.

Let us consider second a cat standing still. During standing still, it has been observed that sometimes there is only soleus force but no gastrocnemius force [14]. Here, one could argue again that soleus is probably recruited as the primary ankle extensor because of its fibre type composition and the corresponding resistance to fatigue. However, the gastrocnemius is a muscle of five to ten times the physiological cross-sectional area of the soleus, and it contains about 25% slow-twitch fibres [3]. Therefore, it is safe to assume that gastrocnemius could satisfy the ankle extensor moment with its slow twitch motor units exclusively, and so be as fatigue-resistant as the soleus. However, aside from ankle extension, the gastrocnemius is also a knee flexor. Therefore, if gastrocnemius were recruited during quiet standing instead of soleus, its (flexor) moment contribution at the knee would have to be compensated for by the knee extensor muscles. This example illustrates that exclusive soleus or gastrocnemius activation during quiet standing can easily be accomplished. However, in contrast to exclusive soleus activation, exclusive gastrocnemius activation requires an increased knee extensor activity (force). Co-contraction of knee flexor and knee extensor muscles requires an increased metabolic cost to provide a zero net knee joint moment, compared to knee extensor activity alone. Also, co-contraction would produce increased knee stiffness compared to knee extensor activity alone. Although measuring hindlimb ground reaction forces and determining hindlimb joint moments would not show a difference for quiet standing performed with the soleus or the gastrocnemius, metabolically and from the point of view of joint stiffness, the two scenarios would be different. Similarly, many other examples could be chosen to illustrate that muscle coordination during movements is functionally not (necessarily) redundant and/or that different patterns of muscle activation resulting in the same joint moments might produce vastly different physiological (energy cost) and mechanical (joint stiffness) conditions.

After this little detour, demonstrating that functionally every movement or limb position might only be possible using a unique set of muscle forces, I would like to come back to the more accepted view that muscles crossing a joint represent a redundant system. As a consequence, it is assumed that a movement can be accomplished using an infinite number of possible muscle interaction patterns. Accepting this view, equations (7.1)–(7.3) are redundant – they contain more unknowns (the magnitudes of all forces) than system equations.

Conceptually, a redundant system of equations may be solved uniquely by decreasing the number of unknowns or increasing the number of system equations until the number of unknowns and the number of equations are equal. Traditionally, decreasing the number of unknowns has been done by grouping agonistic muscles into functional units. Using such an approach, the solutions become the forces in functional groups of muscles rather than the individual muscle forces, which is unacceptable if one is interested in the coordination and movement/moment control of all muscles at a joint.

Increasing the number of system equations until they match the number of unknowns has the advantage that the solutions represent the individual muscle forces. Increasing the number of system equations is typically done by introducing physiological, anatomical, or neural relationships about the system of interest. For example, the relationship between the forces of individual muscles in an agonistic group might be described as a function of the size, fibre type distribution, and fibre length of the muscles involved. This approach has rarely been chosen to obtain predictions of individual muscle forces crossing a joint, probably because it would require the type of insight into the control of synergistic muscles that one tries to obtain with this particular type of research.

In biomechanics, the most frequent approach to solving underdetermined mathematical systems of the type shown in equation (7.1) is optimization. Optimization approaches require the specification of an objective (or cost) function, (equality and or inequality) constraint functions, and design variables. When predicting individual muscle forces during movement, the design variables (typically) are the individual muscle forces, f_i. Constraint functions for individual muscle force predictions typically include the idea that the (active) joint moments are produced by the individual muscle forces (e.g. equations (7.1) or (7.2), and the knowledge that muscle forces are unidirectional (i.e. tensile). Specifying tensile forces as positive, therefore, gives $f_i \geq 0$. Other constraint functions might include upper limits for the muscle forces, constraints requiring synergistic action of agonistic muscles, and inhibiting synergistic action of antagonistic muscles, to name but a few possibilities. Finally, every optimization approach requires the specification of an objective function which contains the design variables. The objective function is then optimized by varying the design variables while obeying all constraint functions until the optimal solution is found.

164 Theoretical models of skeletal muscle

One of the first optimization problems aimed at estimating individual muscle forces in a mathematically redundant system was proposed by Penrod et al. [20]. These researchers suggested that the objective function in a normal, healthy, non-fatigued system should reflect the idea that total muscular effort is minimized:

$$\phi = \rho_1 F_1 + \rho_2 F_2 + \cdots + \rho_n F_n \qquad (7.4)$$

where ϕ, a measure of the total muscular effort, is minimized and ρ_i represent (unspecified) weighing functions. Equation (7.4) was used to determine the forces of seven muscles crossing the wrist joint (with two rotational degrees of freedom).

Penrod et al. [20], like everyone else in the early stages of theory development [4, 22], formulated the problem as a linear optimal design, i.e., the objective function (equation (7.4)) and all (inequality and equality) constraint functions were linear functions of the design variables. Linear optimization was convenient in so far as there were standard solution algorithms available (simplex algorithm, [6]). However, linear optimal designs have severe limitations from a functional point of view; most importantly, they can typically only predict as many non-zero muscle forces as there are independent, non-homogeneous constraint functions (typically, the number of rotational degrees of freedom at the joint of interest). Therefore, Penrod et al. [20] could only predict two non-zero muscle forces among the seven muscles which were included in the problem formulation. Initially, they argued that this finding was consistent with the concept of reciprocal inhibition; however, one of the muscles could not be activated for any of the loading conditions imposed, which appears unrealistic. Based on this last finding, Penrod et al. [20] attempted to increase the number of active muscles by limiting the maximal force that could be exerted by each muscle. Although such a constraint appears theoretically feasible (after all, a muscle has a finite potential to produce force), the idea that a muscle will go from zero activation to being fully activated before a second muscle is recruited is (physiologically) absurd.

In an attempt to overcome the limitations of linear optimal designs, researchers continued to constrain the system in ways so as to enforce synergistic action among agonistic muscles [4]. In these situations, the predicted muscle forces were primarily determined by the constraint functions rather than the objective function. However, it was the objective function which was initially chosen to produce the correct behaviour of the muscles (e.g. minimization of total muscular effort).

Optimization approaches rely on the idea that movement control via the muscles underlies some optimal, physiological criterion. If one accepts that such a physiological criterion exists (at least for a selected group of movements), then standard linear optimal designs relying on constraint functions to produce synergistic actions among agonistic muscles, and enforcing other experimentally observed phenomena, are not useful. If there exists an optimal, physiologic criterion, one should capture it with the objective function.

A significant advance in the theoretical prediction of individual muscle forces was made by Crowninshield and Brand [5], who insisted on the importance of a physiologically based objective function (rather than an arbitrary or mathematically convenient one). These investigators used a nonlinear optimal design formulation. The objective function was based on the idea that during many cyclic, low-level activities (such as normal walking), muscles are recruited in such a way as to maximize endurance (i.e. maximize the duration a movement can be sustained). Based on the well-known, inverse-nonlinear relationship of muscular force and the maximal contraction duration, they specified the objective function as

$$\text{minimize } \phi, \text{ where } \phi = \sqrt[n]{\sum_{i=1}^{m} (f_i/PCSA_i)^n} \qquad (7.5)$$

Equation (7.5) is minimized when the radicand is minimized. The radicand contains the summation of the muscle stresses ($f_i/PCSA_i$) to the nth power, where f_i is the ith muscle force and $PCSA_i$ is the ith physiological cross-sectional area of the muscle.

Using the above objective function (equation (7.5)) in conjunction with the appropriate muscle force ($f_i \geq 0$) and joint moment constraints, Crowninshield and Brand [5] predicted the forces of 47 lower limb muscles during walking. The lower limb joints were considered to have a total of five rotational degrees of freedom (three at the hip, and one each at the knee and ankle, respectively).

Crowninshield and Brand [5] compared their force predictions to the temporal patterns of experimentally determined electromyographical activities of a series of lower limb muscles. They concluded that their force prediction model made good temporal predictions, but were unable to comment on the magnitudes of the predicted muscle forces. Because of the nonlinear character of the optimal design, synergistic activity of agonistic muscles was predicted without enforcement by constraint functions. Also, co-contraction of two-joint antagonistic pairs of muscles was demonstrated, as was co-contraction of two one-joint muscles (soleus and tibialis anterior) crossing the one-degree-of-freedom ankle joint. This latter result is not possible mathematically [10], except if caused by numerical inaccuracies of the applied algorithm. Probably the biggest limitation of Crowninshield and Brand's [5] work was that the predicted muscle force magnitudes could not be validated, leaving the model unproven for a decade, until force predictions were compared to directly measured muscle forces [11].

Validation of theoretically predicted individual muscle forces has remained a problem in this area of research. To this date, there are only few studies which have systematically validated individual muscle force predictions during movements. Dul et al. [7] were the first to do so. These researchers developed a theoretical algorithm for the prediction of individual muscle forces based on a 'minimum-fatigue' criterion. Their algorithm required that the endurance time of

166   Theoretical models of skeletal muscle

each muscle was calculated as a function of its instantaneous force, its maximal force, and its fibre type distribution. Individual muscle force predictions were found by maximizing the endurance time of the muscle with the shortest endurance time. Since the force–endurance relationship depends heavily on the fibre type distribution of the muscle, the individual force predictions also depended on the fibre type distribution.

Dul et al. [7] predicted the forces of the cat medial gastrocnemius (a predominantly fast-twitch fibred muscle) and the cat soleus (a virtually completely slow-twitch fibred muscle). The force sharing between these two muscles was calculated assuming that both muscles' primary function was ankle extension (i.e. any flexor action of the medial gastrocnemius at the knee was ignored). Considering only planar motion (flexion–extension at the ankle), the system of equations had one degree of freedom and two unknowns. The result was an exponential function, whereby the force of the soleus, f_s, could be calculated from the force of the gastrocnemius, f_g, as follows:

$$f_s = 5.08(f_g)^{0.42} \tag{7.6}$$

Inspection of equation (7.6) shows that for small gastrocnemius forces (less than 17 N), the soleus forces are larger than the gastrocnemius forces, whereas the gastrocnemius forces become increasingly more dominant with increasing force demands.

The results predicted by Dul et al. [7], equation (7.6), were compared to the actual measurements of medial gastrocnemius and soleus forces. These comparisons were made for the peak forces achieved by the two muscles during a variety of locomotor conditions including standing still, walking, trotting, galloping, and jumping. The force-sharing predictions (equation (7.6)) were acceptable for the walking and running tasks; however, for jumping the soleus forces were overestimated by a factor of 2 (Figure 7.6).

The limitations of the validation procedures used by Dul et al. [7] included that the researchers did not make their own force measurements but obtained the results from the literature [24]. This fact in itself would not be a big drawback; however, the input data required for the force predictions (maximal muscle forces and fibre type distributions) were not available for the experimental animals, and had to be obtained from yet another literature source [21].

A further limitation of the force prediction procedures by Dul et al. [7] was that the medial gastrocnemius was treated as a single-joint muscle (ankle extensor); its knee flexor function was ignored. Although it is feasible to assume that the gastrocnemius is primarily an ankle extensor, neglecting its knee flexor function resulted in a theoretical model containing just one degree of freedom. It can be shown that by introducing a second degree of freedom (knee flexion–extension), the force sharing between soleus and medial gastrocnemius would no longer be a functional relation (i.e. equation (7.6)) but would become a non-unique relation which depends at any instant in time on both the knee and ankle moment requirements.

Movement control 167

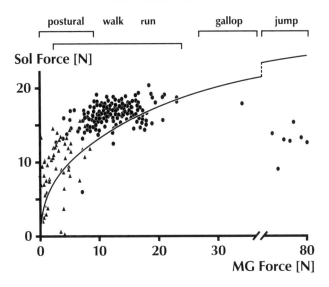

Figure 7.6 Theoretical prediction of the force sharing between cat soleus (Sol) and medial gastrocnemius (MG) for a variety of activities (solid line, Dul *et al.* [7]), and the corresponding peak forces obtained experimentally (triangles, static; dots, dynamic).

A further conceptual problem with the approach of Dul *et al.* (7) was that they only used the peak forces of the two muscles for validation purposes. Aside from ignoring a vast amount of information when just considering the peak forces rather than the entire force–time histories, the peak forces are not appropriate reference values for validation because they do not occur at the same instant in time. By definition of the term 'force sharing' among muscles, the interest lies in comparing the force–time histories of individual muscle forces. During cat locomotion, peak medial gastrocnemius forces systematically precede peak soleus forces; therefore, when using the peak forces for validation of a force-sharing algorithm, values which occur at different instants in time are treated as if they occurred simultaneously.

A further attempt at validating muscle force predictions with experimentally measured *in vivo* muscle forces was made by Herzog and Leonard [11]. In this study, the force predictions of five widely used force-sharing algorithms were compared to directly measured forces in the cat gastrocnemius, soleus, and plantaris for a variety of locomotor activities. Force sharing among the three muscles was calculated analytically for each of the five theoretical models using input from the experimental animal.

The results indicated that none of the theoretical models (including the one proposed in [7]) was able to predict the individual muscle forces in an appropriate way for all locomotor conditions tested. Based on these results, it was concluded that none of the tested movement control models adequately accounted for the force sharing of muscles during voluntary, low-level, cyclic movements.

Since the beginning of the 1990s, little progress has been made in the theoretical prediction of individual muscle forces. A significant amount of new experimental data on force sharing in the cat ankle extensor muscles has been provided [12], but little of the experimental data has been used for the development, testing, or validation of theoretical models of movement control.

Most of the experimental research using *in vivo* multiple muscle force recordings has been performed in the cat ankle extensor and flexor muscles [12, 14, 24]. Specifically, the gastrocnemius, soleus, and plantaris forces have been measured for a variety of activities, primarily locomotion. The results obtained in a variety of experiments performed in different laboratories showed consistent force-sharing behaviour among these three muscles. Some of the conceptual findings which were consistently observed include:

- The force sharing among the cat ankle extensors depends on the activity (here, specifically the speed of locomotion; see Figures 7.7 and 7.8).
- Soleus forces dominate in activities of low force requirements and slow movements (standing, slow walking).
- Plantaris and gastrocnemius forces become dominant in activities of high force demands and fast speeds of movement (running, jumping, scratching, paw shaking).
- Within a range of locomotion speeds from 0.4 m/s to 2.4 m/s, peak forces of the soleus remain about constant, peak forces of the plantaris increase by a factor of about 2–4, and peak gastrocnemius forces increase by a factor of about 4–6 (Figure 7.7).
- The force sharing between the three ankle extensors has a loop shape (Figure 7.8).
- During specific tasks, some of the ankle extensors may not contribute active force while the remaining muscles do. For example, during standing still, gastrocnemius forces may be zero while the soleus is active; during paw shaking soleus forces may be zero while the gastrocnemius is active.

These conceptual results should help to emphasize the tremendous range in force-sharing patterns among agonistic muscles during different tasks. So far, a conceptual framework of movement control which can explain all experimental observations is not available.

A similar list of conceptual observations could also be made for antagonistic pairs of muscles for which *in vivo* force measurements are available. Such data are much rarer than results on agonistic force sharing; however, the behaviour between cat soleus and tibialis anterior has been measured and described. The conceptual results for these two muscles include:

- For slow walking, soleus is active during the stance phase and tibialis anterior is active during the swing phase.

Movement control 169

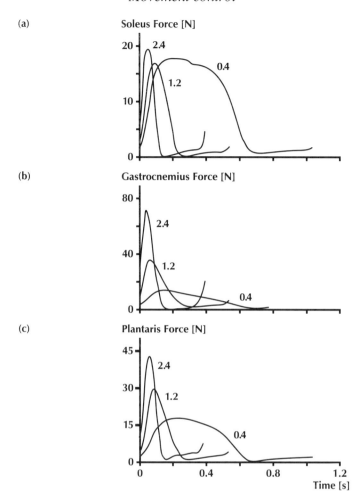

Figure 7.7 Mean force–time histories of cat soleus, gastrocnemius, and plantaris for walking at 0.4 and 1.2 m/s and trotting at 2.4 m/s. Mean values were calculated from a minimum of 10 consecutive step cycles. (Adapted from [7].)

- For fast running, soleus and tibialis anterior remain primarily active during stance and swing, respectively; however, there is a distinct force overlap in the transition from swing to stance (but not from stance to swing).

The first observation corresponds to the expected functional behaviour of the two muscles and is explained by the concept of reciprocal inhibition (the antagonist is inhibited during activity of the agonist and vice versa). The second observation is more interesting than the first one and might be explained in several ways: (1) Early onset of the soleus force (towards the end of swing), which

170 *Theoretical models of skeletal muscle*

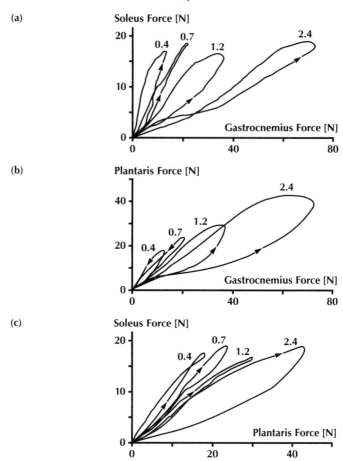

Figure 7.8 Mean force-sharing curves for soleus–gastrocnemius, plantaris–gastrocnemius, and soleus–plantaris for walking at 0.4, 0.7, and 1.2 m/s and trotting at 2.4 m/s. Mean curves were calculated from a minimum of 10 consecutive step cycles.

overlaps with the tibialis anterior force, may be required so that soleus has sufficient time to build up a full active state during the (short) stance phase; (2) as a consequence of the early onset of soleus force, tibialis anterior needs to remain active during the late phase of swing, otherwise there would be a strong (undesired) plantarflexion of the foot prior to paw contact; or (3) co-contraction of the antagonist pair will cause a stiffening of the ankle joint just prior to paw impact, therefore the effects of perturbations (which could not be corrected through feedback at the fast speeds of locomotion) might be minimized. A detailed and universally accepted paradigm for the (stiffness) control of joints through agonist–antagonist co-contraction is not available at present.

7.5 FUTURE RESEARCH IN THE AREA OF MECHANICS IN MOVEMENT CONTROL

Movement control, be it through the control of joint moments or joint stiffness, depends directly on the force contribution of each muscle crossing a joint. Therefore, for a complete understanding of the control of a joint, all individual muscle forces should be known. The maximal number of muscle forces recorded simultaneously at the cat ankle is four; however, there are 12 tendinous units crossing the ankle, leaving eight muscle forces unknown. One of the challenges for the future will be to design force-measuring devices of such versatility that all muscle forces at a joint can be measured simultaneously. All of the existing force transducers are inadequate for such measurements, typically because they are too big.

Many muscles cross more than one joint. For example, from the primary muscles discussed in this chapter (cat gastrocnemius, soleus, plantaris, and tibialis anterior), the gastrocnemius (ankle, knee) and plantaris (ankle, knee, metatarsophalangeal joint via an in-series arrangement with the flexor digitorum brevis) cross more than one joint. Therefore, these muscles influence the moments and the control of several joints. As a consequence, it appears impossible that movement control can be studied completely at a single joint. Simultaneous measurements of individual muscle forces at several neighbouring joints are presently not available, but might produce a wealth of new insight into the control of 'limb' movements. In the cat hindlimb, a combination of simultaneous force recordings from knee and ankle flexor and extensor muscles should be possible without too much difficulty.

One of the severe limitations of research on movement control is that direct *in vivo* force recordings from individual muscles are technically difficult and time-consuming; more importantly, they are virtually impossible to perform in humans. Obviously, if there was a technique to measure (calculate) *in vivo* human muscle forces accurately, completely fresh insight and increased interest in movement control studies might result. Unfortunately, determining muscle forces theoretically using the inverse dynamics approach [2] is more qualitative than quantitative and estimating dynamic muscle forces from electromyographical signals [15–18] has not given accurate results.

One of the ideas on how muscle forces may be coordinated and controlled during movements has been the notion of minimal metabolic cost. According to this idea, synergistic muscles are recruited in such a way that the metabolic cost of movement is minimized. For low-level, repetitive, everyday movements such a 'control' principle is appealing and it has been proposed or used in a variety of forms; but it has never been applied, tested, and validated systematically.

One of the problems associated with minimizing the metabolic cost of movement is how to formulate the metabolic cost of movement in mathematical terms. Hardt [8] derived an explicit formulation of the energy cost of muscular contraction and incorporated this energy function into an optimization scheme aimed at predicting muscle forces in the lower limb during walking. The cost function of the optimization scheme then became: find the muscle forces (for walking) that minimize the energy consumption of the muscles over a given period of time. The results obtained using the minimum energy function were compared to electromyographical signals determined from subjects during walking. Unfortunately, to our knowledge, Hardt's [8] force predictions were not published and no further developments were made from this promising work.

The optimization algorithms proposed by Crowninshield and Brand [5], Dul et al. [7], and Herzog [9] were indirectly aimed at minimizing metabolic cost of movements by either minimizing stress or fatigue, or maximizing endurance of the muscles involved in specific tasks. However, these latter formulations did not contain an explicit relation between the work produced by the muscle and the associated metabolic cost.

In order to predict the forces exerted by individual muscles during movement, the following five step procedure is suggested:

1. A model of the muscle must be defined, i.e. the muscle's force output must be known as a function of contractile conditions, contractile history, and activation.
2. The attachment sites of the muscle within the skeletal system must be defined in order to describe the functional effects of muscle contraction.
3. The target movement, its kinematics, external forces, and the body segments' inertial characteristics must be defined.
4. The control of all muscles defined for the target system must be defined.
5. The predicted force–time histories of the individual muscles must be compared (at least in the model development phase) against the experimentally measured muscle forces for validation of the theoretical model.

Each of these steps comprises its own research area with many challenging problems. Nevertheless, it should be possible to combine these five steps in a collaborative research group, so that a valid control model of individual muscle forces – and therefore movement – might become reality in the near future.

REFERENCES

[1] Abraham, L.D. and Loeb, G.E. (1985) The distal hindlimb musculature of the cat. *Exp. Brain Res.*, **58**: 580–593.
[2] Andrews, J.G. (1974) Biomechanical analysis of human motion. *Kinesiol.*, **4**: 32–42.
[3] Ariano, M.A., Armstrong, R.B. and Edgerton, V.R. (1973) Hindlimb muscle fiber populations of five mammals. *J. Histochem. Cytochem.*, **21(1)**: 51–55.
[4] Crowninshield, R.D. (1978) Use of optimization techniques to predict muscle forces. *J. Biomech. Engng*, **100**: 88–92.
[5] Crowninshield, R.D. and Brand, R.A. (1981) A physiologically based criterion of muscle force prediction in locomotion. *J. Biomech.*, **14**: 793–801.
[6] Dantzig, G.B. (1963) *Linear Programming and Extensions*, Princeton University Press, Princeton, NJ.

[7] Dul, J., Johnson, G.E., Shiavi, R. and Townsend, M.A. (1984) Muscular synergism – II. A minimum-fatigue criterion for load sharing between synergistic muscles. *J. Biomech.*, **17**: 675–684.
[8] Hardt, D.E. (1978) A minimum energy solution for muscle force control during walking. PhD, Thesis, Massachusetts Institute of Technology, Cambridge, MA.
[9] Herzog, W. (1987) Individual muscle force estimations using a non-linear optimal design. *J. Neurosci. Methods*, **21**: 167–179.
[10] Herzog, W. and Binding, P. (1993) Cocontraction of pairs of antagonistic muscles: Analytical solution for planar static nonlinear optimization approaches. *Math. Biosci.*, **118**: 83–95.
[11] Herzog, W. and Leonard, T.R. (1991) Validation of optimization models that estimate the forces exerted by synergistic muscles. *J. Biomech.*, **24**: 31–39.
[12] Herzog, W., Leonard, T.R. and Guimarães, A.C.S. (1993) Forces in gastrocnemius, soleus, and plantaris tendons of the freely moving cat. *J. Biomech.*, **26**: 945–953.
[13] Herzog, W., Zatsiorsky, V., Prilutsky, B.I. and Leonard, T.R.. (1994) Variations in force–time histories of cat gastrocnemius, soleus, and plantaris muscles for consecutive walking steps. *J. Exp. Biol.*, **191**: 19–36.
[14] Hodgson, J.A. (1983) The relationship between soleus and gastrocnemius muscle activity in conscious cats – a model for motor unit recruitment? *J. Physiol.* (London), **337**: 553–562.
[15] Hof, A.L. and Van den Berg, J. (1981) EMG to force processing I: An electrical analogue of the Hill muscle model. *J. Biomech.*, **14**: 747–758.
[16] Hof, A.L. and Van den Berg, J. (1981) EMG to force processing II: Estimation of parameters of the Hill muscle model for the human triceps surae by means of a calf ergometer. *J. Biomech.*, **14**: 759–770.
[17] Hof, A.L. and Van den Berg, J. (1981) EMG to force processing III: Estimation of model parameters for the human triceps surae muscle and assessment of the accuracy by means of a torque plate. *J. Biomech.*, **14**: 771–785.
[18] Hof, A.L. and Van den Berg, J. (1981) EMG to force processing IV: Eccentric–concentric contractions on a spring-flywheel set up. *J. Biomech.*, **14**: 787–792.
[19] Loeb, G.E. (1984) The control and responses of mammalian muscle spindles during normally executed motor tasks. *Exercise Sport Sci. Rev.*, **12**: 157–204.
[20] Penrod, D.D., Davy, D.T. and Singh, D.P. (1974) An optimization approach to tendon force analysis. *J. Biomech.*, **7**: 123–129.
[21] Petrofsky, J.S. and Lind, A.R. (1979) Isometric endurance in fast and slow muscles in the cat. *Am. J. Phsyiol.*, **236**: 185–191.
[22] Seireg, A. and Arvikar, R.J. (1973) A mathematical model for evaluation of force in lower extremities of the musculoskeletal system. *J. Biomech.*, **6**: 313–326.
[23] Smith, J.L., Betts, B., Edgerton, V.R. and Zernicke, R.F. (1980) Rapid ankle extension during paw shakes: Selective recruitment of fast ankle extensors. *J. Neurophysiol.*, **43**: 612–620.
[24] Walmsley, B., Hodgson, J.A. and Burke, R.E. (1978) Forces produced by medial gastrocnemius and soleus muscles during locomotion in freely moving cats. *J. Neurophysiol.*, **41**: 1203–1215.

Appendix A
Topics in time-independent modelling

A.1 SOLVING NONLINEAR PROBLEMS

In muscle mechanics, linearity is the exception rather than the rule. Large displacements, on the one hand, and complex material behaviour, on the other, cause the equilibrium equations, the constitutive relations, and the constraints to conform a system of nonlinear equations, for which there is in general no closed-form solution. Instead, some scheme of numerical approximation must be employed to arrive at a satisfactory solution within a desired range of accuracy. If we assume, for the moment, that there are no time-dependent effects (such as viscosity and inertia), we are led to a system of algebraic (rather than differential) equations which can be written in the following generic form:

$$f_I(x_1, ..., x_N) = 0, \quad I = 1, ..., N \quad (A.1)$$

where $x_1, ..., x_N$ represent N generic unknowns (such as displacement components and Lagrange multipliers) and $f_1, ..., f_N$ denote arbitrary functions (such as might be obtained from equilibrium, kinematic, and/or constraint equations). If each of these functions happens to be linear in each of the unknowns, that is, if

$$f_I = \sum_J a_{IJ} x_J + b_I, \quad I, J = 1, ..., N \quad (A.2)$$

where a_{IJ}, b_I are constants (as was the case in Section 5.10), then we can solve the system exactly by the technique of elimination (**Gauss's method**), which we shall briefly review in Section A.4. If, on the contrary, some or all of the functions f_I are not linear, we may use a variety of iterative techniques of approximation, the most basic of which is the **Newton–Raphson method**. The application of this method requires only that the functions involved be once differentiable. As with all other multi-variable nonlinear schemes, however, convergence to a solution is not unconditionally guaranteed, even when a solution exists.

To get the Newton–Raphson iteration process started, one needs to specify an initial guess for the unknowns. Let $x_1^0, ..., x_N^0$ be the given initial guess. When

Topics in time-independent modelling

these values are plugged into the equations to be solved, Equations (A.1), these equations fail in general to be satisfied, as manifest in the fact that the **residuals**

$$r_I^0 = f_I(x_1^0, ..., x_N^0), \quad I = 1, ..., N \tag{A.3}$$

do not vanish. If, for instance, $f_I = 0$ represents an equilibrium statement, then the residual r_I^0 is an unbalanced force which will cause the structure not to linger at the guessed-at deformed configuration, but to move to a new one. In the method of Newton–Raphson the search for a new (and presumably better) configuration is based on the differential of the functions f_I at the initial guess. Indeed, the very notion of derivative implies a first-order approximation of a nonlinear function by a linear one in the neighbourhood of a given point. Thus we have that the values of f_I at points $x_1, ..., x_N$ in the neighbourhood of $x_1^0, ..., x_N^0$ are, approximately,

$$f_I(x_1, ..., x_N) \approx f_I(x_1^0, ..., x_N^0) + \left[\frac{\partial f_I}{\partial x_1}\right]_0 (x_1 - x_1^0) + \cdots + \left[\frac{\partial f_I}{\partial x_N}\right]_0 (x_N - x_N^0) \tag{A.4}$$

where $[\partial f_I/\partial x_J]_0$ denotes the partial derivative of f_I with respect to x_J evaluated at the initial guess $x_1^0, ..., x_N^0$. This expression can be written more compactly as:

$$f_I(x_1, ..., x_N) \approx r_I^0 + \sum_J \left[\frac{\partial f_I}{\partial x_J}\right]_0 (x_J - x_J^0), \quad J = 1, ..., N \tag{A.5}$$

where equation (A.3) has been invoked. We note that these expressions are of the form (A.2), that is, they are linear in the unknowns x_J, since the derivatives are evaluated at a given fixed point. Equation (A.5) provides us then with a linear estimate of the value of f_I, namely, with a linear approximation to the left-hand side of our system (A.1). Setting this estimate to zero, we obtain the system of linear equations:

$$\sum_J \left[\frac{\partial f_I}{\partial x_J}\right]_0 \Delta x_J = -r_I^0, \quad I, J = 1, ..., N \tag{A.6}$$

for the N increments $\Delta x_J = (x_J - x_J^0)$. The coefficients of this system are the partial derivatives at the initial guess, and the right-hand side is equal to minus the initial ('unbalanced') residuals.

Adopting now the new values

$$x_J^1 = x_J^0 + \Delta x_J, \quad J = 1, ..., N \tag{A.7}$$

as a new guess, the procedure can be repeated as many times as necessary to bring the residuals under a prespecified range of error. As convergence criterion one may use either the magnitude of the residuals or, better still, a measure of the relative stabilization of the successive guesses. One possible indication that the solution is becoming stable is that the increments obtained from the solution of the linear system (A.4) are very small compared with the new values of the unknowns. More precisely, this can be measured in terms of the relative

Appendix A

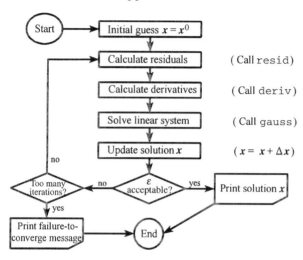

Figure A.1 Flow chart.

quadratic error:

$$\varepsilon^S = \frac{[\sum_J (x_J^S - x_J^{S-1})^2]^{1/2}}{[\sum_J (x_J^S)^2]^{1/2}}, \qquad J = 1,...,N \tag{A.8}$$

in which S denotes the iteration step ($S = 1, ...$).

It is worthwhile noting that at each step of the procedure one needs to calculate not only the N new residuals x_J^S, but also the N^2 derivatives $[\partial f_I/\partial x_J]_S$. This fact renders the method somewhat computationally expensive, particularly for very large systems of equations.

To summarize the Newton–Raphson scheme and also to lay down the groundwork for its computer implementation, we present a sketchy flow chart (Figure A.1). Subroutines to calculate residuals (RESID) and derivatives (DERIV) need to be supplied externally, as well as a linear solver (GAUSS).

A.2 A CODE FOR THE NEWTON–RAPHSON TECHNIQUE

We now produce a computer code in QBASIC for the Newton–Raphson solution of a system of nonlinear equations, assuming that functions RESID, DERIV and subroutine GAUSS are conveniently available.

```
SUB newrap (max%, eps, h, n%, x0())
'
'*******************************************************************************
' Newton-Raphson nonlinear solver
' max% = maximum number of iterations
' eps = tolerance
```

Topics in time-independent modelling 177

```
' h = increment for (possibly) numerical differentiation (see FUNCTION deriv)
' n% = number of equations and unknowns
' x0() = initial and updated solution
' deltax() = Newton-Raphson-generated increments of x0()
' ****************************************************************************
DIM coef(n%, n%), rhs(n%), deltax(n%)
FOR k% = 1 TO max%
' Prepare right-hand side vector rhs() and coefficient matrix coef()
  FOR i% = 1 TO n%
    rhs(i%) = -resid(i%, x0(), n%)
    FOR j% = 1 TO n%
      coef(i%, j%) = deriv(i%, j%, -rhs(i%), x0(), n%, h)
    NEXT j%
  NEXT i%
' Solve system of linear equations
  CALL gauss(coef(), rhs(), deltax(), n%)
' Update solution and calculate quadratic error
  sum1 = 0!
  sum2 = 0!
  FOR i% = 1 TO n%
    x0(i%) = x0(i%) + deltax(i%)
    sum1 = sum1 + x0(i%) * x0(i%)
    sum2 = sum2 + deltax(i%) * deltax(i%)
  NEXT i%
  IF (SQR(sum2 / sum1) <= eps) THEN GOTO 99
NEXT k%
PRINT "Convergence not achieved in Newrap after "; max%; "iterations"
STOP
99 ' Convergence achieved
END SUB
```

A.3 OBTAINING THE RESIDUALS AND DERIVATIVES DIRECTLY FROM VIRTUAL WORK

In this section we produce a computer code to calculate the residuals and derivatives needed for Newton–Raphson directly from the principle of virtual work.

Although computationally speaking it is best to have the governing equations (A.1) available in as explicit a manner as possible, from the analyst's point of view it is convenient also to be able to program the whole procedure using just a single master equation, namely, the virtual work identity (5.60). The main disadvantage of this approach is that many operations are superfluously repeated and many zeros recalculated. On the other hand, the simplicity of the coding justifies the approach from the standpoint of both expediency and conceptual elegance and generality. For practical applications, the reader is encouraged to develop more efficient, albeit less flexible, programs. All computer programs presented in this chapter are driven by pedagogical, rather than computational, objectives.

Let a function `vw(x(), dx(), n%)` be available which provides the value of the total virtual work (internal − external − constraint) for all possible values of the variables $x()$ and their variations $dx()$. Later on we will exhibit such a routine explicitly. Then, a procedure to calculate the residual of the Ith equation may be obtained by simply calculating the virtual work when all variations are zero except the Ith one, which is set to 1. A program to do just that is the following:

```
FUNCTION resid (i%, x(), n%)
DIM dx(n%)
FOR k% = 1 TO n%
dx(k%) = 0!
NEXT k%
dx(i%) = 1!
resid = vw(x(), dx(), n%)
END FUNCTION
```

As far as concerns the derivatives $[\partial f_I/\partial x_J]$ needed for Newton–Raphson's procedure, it is often very difficult to obtain them explicitly, but one can always resort to numerical differentiation. All that is needed is the value of the function f_I at two points, as per the definition of partial derivative:

$$\left[\frac{\partial f_I}{\partial x_J}\right]_0 = [f_I(x_1^0, ..., x_J^0 + h, ..., x_N^0) - f_I(x_1^0, ..., x_J^0, ..., x_N^0)]/h \qquad (A.9)$$

where h is a conveniently chosen small increment. The use of this formula instead of the exact derivatives may be somewhat detrimental to the convergence of the Newton–Raphson scheme. But we note that if the convergence criterion is eventually fulfilled, the quality of the solution is independent of the method used to calculate the derivatives. At any rate, the approximate formula permits us to obtain good estimates of the derivatives with just one extra call to `resid` per derivative, i.e. a total of N^2 extra calls per iteration step. A possible code is:

```
FUNCTION deriv (i%, j%, fi0, x(), n%, h)
' ********************************************************************************
' This function calculates a derivative numerically by forward differences.
' To save computing effort, it is assumed that the value of the function at
' the original point has already been calculated: it is passed as argument
' fi0.h is a given non-zero increment. Ideally, this h could be varied
' for different kinds of variables.
' ********************************************************************************
xx = x(j%)
x(j%) = x(j%) + h
fi0h = resid(i%, x(), n%)
x(j%) = xx
deriv = (fi0h - fi0) / h
END FUNCTION
```

A.4 A LINEAR SOLVER

In this section we write a computer code to solve a system of simultaneous linear equations. The generic system of N linear equations (A.2) can be written explicitly as:

$$a_{11}x_1 + a_{12}x_2 + \cdots + a_{1N}x_N = -b_1 \quad \text{(A.10)}$$

$$a_{21}x_1 + a_{22}x_2 + \cdots + a_{2N}x_N = -b_2 \quad \text{(A.11)}$$

$$\cdots$$

$$a_{N1}x_1 + a_{N2}x_2 + \cdots + a_{NN}x_N = -b_N \quad \text{(A.12)}$$

The key idea behind the *Gauss elimination procedure* for solving this system exactly is that the solution of a linear system is unchanged under the following operations: multiplication of an equation by a non-zero number; and replacement of an equation by the sum of itself plus any other equation of the system. It follows that any convenient combinations of the above operations may be used to obtain an equivalent system whose solution can be obtained more easily than that of the original system. In particular, if all the coefficients on the left-hand side of one of the equations vanish, except one, the value of the corresponding unknown is obtained directly. To effect such a simplification, we start by replacing the second equation by the following admissible combination: itself minus a_{21}/a_{11} times the first equation, i.e.

$$0 \cdot x_1 + \left[a_{22} - \frac{a_{21}}{a_{11}}a_{12}\right]x_2 + \cdots + \left[a_{2N} - \frac{a_{21}}{a_{11}}a_{1N}\right]x_N = -b_2 + \frac{a_{21}}{a_{11}}b_1 \quad \text{(A.13)}$$

After a similar modification is effected on the remaining equations, the system becomes:

$$a_{11}x_1 + a_{12}x_2 + \cdots + a_{1N}x_N = -b_1 \quad \text{(A.14)}$$

$$0 \cdot x_1 + \left[a_{22} - \frac{a_{21}}{a_{11}}a_{12}\right]x_2 + \cdots + \left[a_{2N} - \frac{a_{21}}{a_{11}}a_{1N}\right]x_N = -b_2 + \frac{a_{21}}{a_{11}}b_1 \quad \text{(A.15)}$$

$$\cdots$$

$$0 \cdot x_1 + \left[a_{N2} - \frac{a_{N1}}{a_{11}}a_{12}\right]x_2 + \cdots + \left[a_{NN} - \frac{a_{N1}}{a_{11}}a_{1N}\right]x_N = -b_N + \frac{a_{N1}}{a_{11}}b_1 \quad \text{(A.16)}$$

This system has the same solution as the original one. Leaving now the first equation untouched and considering the remaining $N - 1$ equations, which do not involve x_1 at all, we are now in a position of repeating the previous procedure for this new subsystem, and so on. After $N - 1$ iterations of this procedure, we obtain an equivalent system of the 'triangular' form:

$$a_{11}x_1 + a_{12}x_2 + \cdots + a_{1N}x_N = -b_1 \quad \text{(A.17)}$$

$$A_{22}x_2 + \cdots + a_{2N}x_N = -B_2 \quad \text{(A.18)}$$

$$\cdots$$

$$A_{NN}x_N = -B_N \quad \text{(A.19)}$$

with new constants A_{IJ}, B_I arising from the repeated application of the procedure. The solution of this new system, which must be the same as that of the original system, is easily obtained in domino-like fashion starting from the bottom up. Indeed, the last equation (A.19) already gives us:

$$x_N = -\frac{B_N}{A_{NN}} \quad \text{(A.20)}$$

Similarly, from the penultimate equation of the modified system we may read off:

$$x_{N-1} = \frac{-B_{N-1} - A_{N-1,N} x_N}{A_{N-1,N-1}} \quad \text{(A.21)}$$

and so on.

We have tacitly assumed that at all stages of the procedure the division by the coefficient A_{II} of x_I in the Ith equation was possible. If it should happen that this coefficient is zero, all that needs to be done is to adopt another coefficient as 'pivot', either by renaming the unknowns or changing the order of the equations. If the search for a new non-zero pivot proves to be unsuccessful, the system is said to be **singular**: it either has no solution or an infinite number of them. Singularity is often an indication of inadmissible physical behaviour (e.g. zero stiffness in some direction, improper support, etc.) or of an error of assembly (such as insufficient conditions of support).

In many applications, the sparse interconnectivity of the nodes results in a coefficient matrix which is 'banded', that is, all of the coefficients away from a more or less narrow band around the main diagonal ($a_{11}, a_{22}, ..., a_{NN}$) are already zero. The elimination procedure can, therefore, be significantly shortened by using this information efficiently. However, in keeping with our mainly pedagogical spirit, we do not engage in such refinements, but present here a standard Gauss elimination procedure.

```
SUB gauss (coef(), rhs(), soln(), n%)
DIM aux(n%, n%)
FOR i% = 1 TO n%
soln(i%) = rhs(i%)
FOR j% = 1 TO n%
aux(i%, j%) = coef(i%, j%)
NEXT j%
NEXT i%
IF n% = 1 THEN GOTO 58
FOR k% = 2 TO n%
IF aux(k% - 1, k% - 1) <> 0! THEN GOTO 57
FOR l% = k% TO n%
IF aux(l%, k% - 1) = 0! THEN GOTO 55
FOR m% = k% - 1 TO n%
x = aux(k% - 1, m%)
aux(k% - 1, m%) = aux(l%, m%)
```

```
aux(l%, m%) = x
NEXT m%
x = soln(k% - 1)
soln(k% - 1) = soln(l%)
soln(l%) = x
GOTO 57
55 ' keep looking for non = zero pivot
NEXT l%
56 PRINT "singularity detected by Gauss"
STOP
57 ' successful non-zero pivot search. Equations exchanged if necessary.
FOR i% = k% TO n%
x = aux(i%, k% - 1) / aux(k% - 1, k% - 1)
soln(i%) = soln(i%) - x * soln(k% - 1)
FOR j% = k% TO n%
aux(i%, j%) = aux(i%, j%) - x * aux(k% - 1, j%)
NEXT j%
NEXT i%
NEXT k%
58 IF aux(n%, n%) = 0! THEN GOTO 56
soln(n%) = soln(n%) / aux(n%, n%)
IF n% = 1 THEN GOTO 59
FOR k% = n% - 1 TO 1 STEP -1
FOR j% = k% + 1 TO n%
soln(k%) = soln(k%) - aux(k%, j%) * soln(j%)
NEXT j%
soln(k%) = soln(k%) / aux(k%, k%)
NEXT k%
59 ' Solution completed
END SUB
```

A.5 CODING THE VIRTUAL WORK

In this section we present a computer routine to calculate the total virtual work expression:

$$IVW - EVW \equiv \sum_\alpha \delta(\lambda_\alpha \phi_\alpha), \quad \alpha = 1, ..., p \qquad (A.22)$$

for a muscle-like planar assembly of the particular form shown in Figure A.2. We assume panel area preservation. The total number of nodes M is arbitrary, but the numbering is assumed to follow the pattern shown in the figure. A single force F is applied at the last node in the x-direction. It is assumed that a function cons, with appropriate arguments, is available to calculate the internal forces in the bars according to some given constitutive equations. This function may be used in conjunction with the previous programs to calculate the residuals of the nonlinear equations. It should be emphasized once more that this method of calculating those residuals is extremely inefficient, but it serves the purpose of illustrating how the principle of virtual work contains all the information necessary for the formulation and solution of complex problems. It is not difficult to see how this procedure may be improved, mainly by avoiding the calculation of the virtual work in bars which are known a priori to produce no virtual work for a given variation. Thus, for instance, if the only non-vanishing virtual displacement corresponds to node 2, it is clear that only bars 1–2 and 2–4 will produce internal virtual work. A similar remark applies to the constraints. Using these hints, the reader should be able to produce a program that reduces the computing effort by at least one order of magnitude. Here we present the most naïve version. To make matters even less sophisticated, the support constraints are introduced by means of Lagrange multipliers, an obvious waste of effort and an unnecessary loss of accuracy.

To avoid passing parameters from one routine to another, a COMMON SHARED statement is assumed to be available in the main program, containing the initial geometry and the value of the force applied. With this routine in hand, for all its computational inefficiency, combined with the other routines presented in previous examples, we have almost completed the task of assembling a fully fledged program for the solution of complex muscle models: we only need to provide the constitutive law (FUNCTION cons) and a main program to handle the input and the output. With the addition of those two elements we will have a complete muscle model which is fully nonlinear and which can accomodate any time-independent constitutive behaviour. The neural activation effect can be included in the constitutive law, as will be shown later.

```
FUNCTION vw (u(), du(), n%)
' Calculates the total virtual work of the structure of Fig. 6.18.
' All constraints, including support conditions, are treated with
' Lagrange multipliers. Panel area preservation is enforced.
' The following variables are available through COMMON:
'    mc% = total number of nodes
'    xc(i%,k%) = kth component of initial position of ith node
```

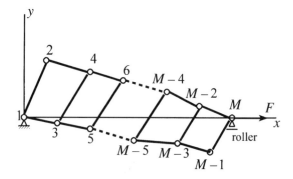

Figure A.2 Generic structure for Section A.5.

Topics in time-independent modelling 183

```
'    forcec = external force F
' (note that n% = 5*mc% - 2 + 2)
sum = 0!
FOR i% = 1 TO mc% - 1
' save some time by using only adjacent bars
mfar% = i% + 2
IF mfar% > mc% THEN mfar% = mc%
kpar% = i% - 2 * INT((i% - 1) / 2)
FOR j% = i% + kpar% TO mfar%
aux = 0!
oldl = 0!
newl = 0!
FOR k% = 1 TO 2
dij0 = xc(j%, k%) - xc(i%, k%)
dij = dij0 + u(2 * j% - 2 + k%) - u(2 * i% - 2 + k%)
oldl = oldl + dij0 * dij0
newl = newl + dij * dij
aux = aux + dij * (du(2 * j% - 2 + k%) - du(2 * i% - 2 + k%))
NEXT k%
oldl = SQR(oldl)
newl = SQR(newl)
sum = sum + cons(i%, j%, oldl, newl) * aux / newl
NEXT j%
NEXT i%
' subtract external virtual work
sum = sum - forcec * du(2 * mc% - 1)
' add panel area conservation
np% = mc% / 2 - 1
FOR i% = 1 TO np%
' calculate areas as cross-products of panel diagonals
v1x = xc(2 * i% + 2, 1) - xc(2 * i% - 1, 1)
v1y = xc(2 * i% + 2, 2) - xc(2 * i% - 1, 2)
v2x = xc(2 * i%, 1) - xc(2 * i% + 1, 1)
v2y = xc(2 * i%, 2) - xc(2 * i% + 1, 2)
area = .5 * (v1x * v2y - v1y * v2x)
v1x = v1x + u(4 * i% + 3) - u(4 * i% - 3)
v1y = v1y + u(4 * i% + 4) - u(4 * i% - 2)
v2x = v2x + u(4 * i% - 1) - u(4 * i% + 1)
v2y = v2y + u(4 * i%) - u(4 * i% + 2)
area1 = .5 * (v1x * v2y - v1y * v2x)
sum = sum + (area1 - area) * du(2 * mc% + i%)
dv1x = du(4 * i% + 3) - du(4 * i% - 3)
dv1y = du(4 * i% + 4) - du(4 * i% - 2)
dv2x = du(4 * i% - 1) - du(4 * i% + 1)
dv2y = du(4 * i%) - du(4 * i% + 2)
darea1 = .5 * (v1x * dv2y + dv1x * v2y - v1y * dv2x - dv1y * v2x)
sum = sum + u(2 * mc% + i%) * darea1
NEXT i%
```

184   Appendix A

```
' Add nodal supports
sum = sum + du(1) * u(2 * mc% + np% + 1) + u(1) * du(2 * mc% + np% + 1)
sum = sum + du(2) * u(2 * mc% + np% + 2) + u(2) * du(2 * mc% + np% + 2)
sum = sum + du(2 * mc%) * u(2 * mc% + np% + 3)
sum = sum + u(2 * mc%) * du(2 * mc% + np% + 3)
vw = sum
END FUNCTION
```

A.6 CODING THE EQUILIBRIUM AND CONSTRAINT EQUATIONS DIRECTLY

For the planar muscle model of Figure A.2, we now write a computer routine to program directly the equations of equilibrium and the panel area constraints representing incompressibility.

In the previous section we demonstrated how the principle of virtual work itself can be used as a numerical tool for the generation of residuals, even when explicit equations are not available. For a straight-line assembly, however, we were able to use the principle of virtual work as a theoretical tool to obtain explicit expressions for the equilibrium equations, equation (5.55), and we have shown how to include the influence of the constraints in equation (5.61). We should, therefore, be able to program those equations directly. Both procedures should yield identical results. When assembling a total program, we will thus have a choice of using our previous function resid in combination with vw, or, alternatively, to use the following version of resid, which should be numerically more efficient. For the sake of variety, however, we have decided in this case not to introduce the supports through Lagrange multipliers, but rather to enforce them by substituting the conditions of vanishing displacement components for the corresponding equilibrium equations. This implies that the number of unknowns is now three less than before. Accordingly, in the main program eventually to be used in preparation of the parameters, this version of resid will require n% =5*mc%/2 - 1, as opposed to n%=5*mc%/2 + 2 for the old version, where mc% is the total number of nodes and n% is the total number of unknowns.

The computing time is greatly reduced by using this new version of resid. Nevertheless, the program is still not as efficient as it could be. For example, the function deriv could be significantly improved by a reasoned a priori recognition of the vanishing mutual stiffnesses between non-contiguous degrees of freedom. The programming of the influence of the panel area constraints is purposely somewhat clumsy, for the sake of transparency. The choice of panel area conservation in no way implies that it is the most realistic or desirable constraint in muscle models, but is general enough to illustrate in a non-trivial manner the method for handling any other geometrical constraint. It should also be noted that as an attractive alternative to the use of constraints one could model some kind of 'connective tissue' with, say, elastic properties (e.g. pressure proportional to area of panel). The addition of the corresponding term to the internal

Topics in time-independent modelling

virtual work will then completely replace the Lagrange multipliers and convey essentially the same physical meaning provided the tissue stiffness becomes very large.

```
FUNCTION resid (i%, x(), n%)
' Available through COMMON SHARED:
'                 mc% = number of nodes
'                 forcec = applied force
'                 xc(j%,k%) = initial kth coord of jth node
' Support conditions enforced directly through equations u = 0.
' Only panel constraints are treated by means of Lagrange multipliers.
' NOTE: For this version of RESID, the main program should specify n% = 5*mc%/2-1
IF i% > 2 GOTO 11
sum = x(i%)
GOTO 49
11 IF i% >= 2 * mc% GOTO 44
sum = 0!
'number of panels
np% = mc% / 2 - 1
'find left-panel number
pn% = INT((i% - 1) / 4)
pnl% = pn%
pnr% = pn% + 1
IF pnl% = 0 THEN pnl% = pnr%
IF pnl% = np% THEN pnr% = pnl%
'influence of panels on node (derivatives of constraint equations times multiplier)
FOR jp% = pnl% TO pnr%
v1x = xc(2 * jp% + 2, 1) - xc(2 * jp% - 1, 1)
v1y = xc(2 * jp% + 2, 2) - xc(2 * jp% - 1, 2)
v2x = xc(2 * jp%, 1) - xc(2 * jp% + 1, 1)
v2y = xc(2 * jp%, 2) - xc(2 * jp% + 1, 2)
v1x = v1x + x(4 * jp% + 3) - x(4 * jp% - 3)
v1y = v1y + x(4 * jp% + 4) - x(4 * jp% - 2)
v2x = v2x + x(4 * jp% - 1) - x(4 * jp% + 1)
v2y = v2y + x(4 * jp%) - x(4 * jp% + 2)
ON (i% - 4 * jp% + 4) GOTO 31, 32, 33, 34, 35, 36, 37, 38
31 sum = sum + .5 * (-v2y) * x(2 * mc% + jp%)
GOTO 39
32 sum = sum + .5 * (v2x) * x(2 * mc% + jp%)
GOTO 39
33 sum = sum + .5 * (-v1y) * x(2 * mc% + jp%)
GOTO 39
34 sum = sum + .5 * (v1x) * x(2 * mc% + jp%)
GOTO 39
35 sum = sum + .5 * (v1y) * x(2 * mc% + jp%)
GOTO 39
36 sum = sum + .5 * (-v1x) * x(2 * mc% + jp%)
```

```
GOTO 39
37 sum = sum + .5 * (v2y) * x(2 * mc% + jp%)
GOTO 39
38 sum = sum + .5 * (-v2x) * x(2 * mc% + jp%)
39 '
NEXT jp%
'equilibrium influences
'find node number
nn% = INT((i% + 1) / 2)
'find parity
k% = i% - 2 * INT((i% - 1) / 2)
FOR jn% = nn% - 2 TO nn% + 2 STEP 2
IF jn% <= 0 OR jn% >= mc% + 1 THEN GOTO 43
jnn% = jn%
IF jn% <> nn% GOTO 41
jnn% = jn% + 1
IF nn% = 2 * INT(nn% / 2) THEN jnn% = jn% - 1
41 aux = 0!
oldl = 0!
newl = 0!
FOR ks% = 1 TO 2
dij0 = xc(jnn%, ks%) - xc(nn%, ks%)
dij = dij0 + x(2 * jnn% - 2 + ks%) - x(2 * nn% - 2 + ks%)
oldl = oldl + dij0 * dij0
newl = newl + dij * dij
IF ks% = k% THEN aux = aux + dij
NEXT ks%
oldl = SQR(oldl)
newl = SQR(newl)
sum = sum - cons(nn%, jnn%, oldl, newl) * aux / newl
43 NEXT jn%
IF i% = 2 * mc% - 1 THEN sum = sum - forcec
GOTO 49
44 IF i%>2 * mc% GOTO 46
sum = x(i%)
GOTO 49
46 ' panel constraint equations
'find panel number
pn% = i% - 2 * mc%
' calculate panel areas as cross-products of panel diagonals
v1x = xc(2 * pn% + 2, 1) - xc(2 * pn% - 1, 1)
v1y = xc(2 * pn% + 2, 2) - xc(2 * pn% - 1, 2)
v2x = xc(2 * pn%, 1) - xc(2 * pn% + 1, 1)
v2y = xc(2 * pn%, 2) - xc(2 * pn% + 1, 2)
area = .5 * (v1x * v2y - v1y * v2x)
v1x = v1x + x(4 * pn% + 3) - x(4 * pn% - 3)
v1y = v1y + x(4 * pn% + 4) - x(4 * pn% - 2)
```

```
v2x = v2x + x(4 * pn% - 1) - x(4 * pn% + 1)
v2y = v2y + x(4 * pn%) - x(4 * pn% + 2)
area1 = .5 * (v1x * v2y - v1y * v2x)
sum = area1 - area
49 resid = sum
END FUNCTION
```

A.7 PUTTING IT ALL TOGETHER: A PROGRAM FOR STATIC ANALYSIS OF SKELETAL MUSCLE

As already pointed out, the only missing essential element for the completion of a fully fledged program for the analysis of a muscle model is a routine containing the constitutive law. The constitutive assumptions embodied in equations (6.1)–(6.4) are adopted here for the sake of illustration. The following computer code implements these constitutive laws in the new function cons. Some more or less straightforward modifications have been effected in other routines to accommodate a tendon attached to the last node of the muscle, and to permit the options of either displacement-controlled or force-controlled deformations. The interactive input makes the program virtually self explanatory. The output is written onto the file STATIC.OUT. The complete listing of the program follows and a detailed example is provided in the next section.

```
DEFDBL A-Z
DECLARE FUNCTION resid# (i%, x#(), n%)
DECLARE FUNCTION cons# (i%, j%, old1#, new1#)
DECLARE SUB newrap (max%, eps#, h#, n%, x0#())
DECLARE FUNCTION deriv# (i%, j%, fi0#, x#(), n%, h#)
DECLARE SUB gauss (coef#(), rhs#(), soln#(), n%)
COMMON SHARED np%, mc%, xc(), forcec, ynf$, fibre(), bot(), top(), tendon, activ(),lr0
OPEN "c:static.out" FOR OUTPUT AS #1
PRINT "This is a program for the static solution of a straight-line planar"
PRINT " muscle model. Follow the self-explanatory input instructions."
PRINT " "
PRINT "              GENERAL APPEARANCE"
PRINT " "
PRINT " y-axis                o:node "
PRINT "    ^      2           f:fibre"
PRINT "    |   o     4        a:aponeurosis "
PRINT "    |  /    a   o"
PRINT "    |  /      /  a    6"
PRINT "    |  f/   f/     o"
PRINT "    | / panel /   f/  ."
PRINT "    | /  # 1  / panel /    .   2p           2p + 3"
PRINT "    | /      /   #2  /    o   2p+2        (roller & force"
PRINT "    |/      /       /    f/   a (roller)    or fixed)"
PRINT "    o-------/--------/----------------/---------o-------------o------- >"
PRINT "    1  a   o     /            / panel /    tendon          x-axis"
```

```
PRINT " (fixed)   3      a o                /  # p /f"
PRINT "                  5   '      /         /"
PRINT "                           o     a     o"
PRINT "                                2p-1    2p+1"
INPUT "enter number of panels : ", np%
' calculate total number of nodes
mc% = 2 * np% + 3
DIM xc(mc%, 2), fibre(np% + 1), bot(np%), top(np%), activ(np% + 1)
PRINT "next, you will enter the nodal coordinates in the x-y system depicted"
PRINT"                       (in mm)"
FOR i% = 1 TO mc%
PRINT "for node # ", i%; : INPUT ; " x = ", xc(i%, 1): INPUT " y = ", xc(i%, 2)
NEXT i%
PRINT "is the last node fixed, i.e., displacement controlled (y),"
INPUT "or is it movable, i.e., force controlled (n)? Enter y or n: ", ynf$
'calculate total number of unknowns
n% = 5 * (mc% - 1) / 2 + 1
PRINT "next you will enter the fibre-, aponeuroses- and tendon-areas of influence"
PRINT "                       (in sq mm)"
FOR i% = 1 TO np%
PRINT "fibre "; 2 * i% - 1; "-"; 2 * i%;: INPUT; ": ", fibre(i%): PRINT " apo. "; 2 * i% -
1; "-"; 2 * i% + 1; : INPUT ; ": ", bot(i%): PRINT " apo. "; 2 * i%; "-"; 2 * i% + 2; : INPUT
": ", top(i%)
NEXT i%
PRINT "fibre "; 2 * np% + 1; "-"; 2 * np% + 2; : INPUT ; ": ", fibre(np% + 1): INPUT "
tendon: ", tendon
PRINT "next, you will enter the ratio initial length to optimal length"
INPUT ; "(This ratio is assumed to be the same for all fibres) lr0 = ", lr0
PRINT "next, you will enter parameters to control the numerical procedure:"
PRINT " max = maximum number of iterations in Newton-Raphson (say, 50)"
PRINT " eps = relative quadratic error for convergence (say, 0.005)"
PRINT " h   = increment for numerical differentiation (say, 0.001)"
INPUT ; "max = ", max%: INPUT ; " eps = ", eps: INPUT" h =", h
' set initial conditions to zero for first given loading
DIM x0(n%)
FOR i% = 1 TO n%
x0(i%) = 0#
NEXT i%
PRINT "Finally, you will enter the loading conditions"
77 '
IF ynf$ <> "n" THEN GOTO 81
INPUT "You have specified a movable end. Therefore, enter the force (in N):", forcec
GOTO 82
81 INPUT "You have specified controlled displacement. Enter it (in mm):", forcec
82 PRINT "You will now enter the fibre activation coefficients (between 0 and 1)"
FOR i% = 1 TO np% + 1
PRINT "fibre "; 2 * i% - 1; "-"; 2 * i%; : INPUT ": ", activ(i%)
NEXT i%
CALL newrap(max%, eps, h, n%, x0())
```

Topics in time-independent modelling

```
PRINT #1, " "
PRINT #1, "solution for applied"; forcec
PRINT #1, "NODE          X-DISPL          Y-DISPL"
FOR i% = 1 TO mc%
PRINT #1, i%, x0(2 * i% - 1), x0(2 * i%)
NEXT i%
' calculate internal forces
'
'
PRINT #1, " "
PRINT #1, "FIBRE   FORCE   APO   FORCE   APO   FORCE"
FOR ip% = 1 TO np% + 1
nn% = 2 * ip% - 1
jnn% = 2 * ip%
FOR ic% = 1 TO 3
oldl = 0#
newl = 0#
FOR ks% = 1 TO 2
dij0 = xc(jnn%, ks%) - xc(nn%, ks%)
dij = dij0 + x0(2 * jnn% - 2 + ks%) - x0(2 * nn% - 2 + ks%)
oldl = oldl + dij0 * dij0
newl = newl + dij * dij
NEXT ks%
oldl = SQR(oldl)
newl = SQR(newl)
fff = cons(nn%, jnn%, oldl, newl)
PRINT #1, nn%; jnn%; fff;
IF ip% = np% + 1 THEN GOTO 84
jnn% = jnn% + 1
IF jnn% = 2 * (ip% + 1) THEN nn% = nn% + 1
NEXT ic%
PRINT #1, " "
NEXT ip%
84 '
'
oldl = xc(mc%, 1) - xc(mc% - 1, 1)
newl = oldl + x0(2 * mc% - 1) - x0(2 * mc% - 3)
oldl = ABS(oldl)
newl = ABS(newl)
PRINT #1, "calculated tendon force = ", cons(mc% - 1, mc%, oldl, newl)
'update for next run
PRINT " continue? y or n": INPUT yn$: IF yn$ <> "y" THEN GOTO 88
PRINT "You have decided to continue. Present values will be used as initial"
PRINT "You will now enter the new loading. Follow instructions as before"
GOTO 77
88 '
END


FUNCTION cons (i%, j%, oldl, newl)
' Available through COMMON SHARED:
```

Appendix A

```
'np%, mc%, xc(), forcec, ynf$, fibre(), bot(), top(), tendon, activ(), dt, mass(), lr0
aux = 0!
' distinguish between fibres and aponeuroses (or tendon)
big% = i%
small% = j%
IF small%<big% GOTO 109
big% = j%
small% = i%
109 dif% = big% - small%
ON (dif% + 1) GOTO 117, 111, 115, 117
111 IF big% = mc% THEN GOTO 114
IF small% = 2 * INT(small% / 2) THEN GOTO 117
' fibre constitutive equation
optl = oldl / lr0
lratio = newl / optl
'force-length influence
fl = -.772 * lr0 ^ 2 + 1.544 * lr0 - .494
aux = activ(big% / 2) * fibre(big% / 2) * (fl + 0.9664 * (lratio - lr0))
GOTO 117
114 area = tendon
GOTO 116
115 panel% = INT((big% - 1) / 2)
area = bot(panel%)
IF big% = 2 * (panel% + 1) THEN area = top(panel%)
' tendon-like constitutive equation
116 strain = (newl - oldl) / oldl
' force in Newtons when area in square mm
aux = (15160! * ((strain + .000725) ^ 2 - 5.26E-07)) * area
' extend linearly to the (unphysical) compressive range
IF strain<0 THEN aux = 15160! * 2! *.000725 * strain * area
117 cons = aux
END FUNCTION


FUNCTION deriv (i%, j%, fi0, x(), n%, h)
' ************************************************************************************
' This function calculates a derivative numerically by forward differences.
' To save computing effort, it is assumed that the value of the function at
' the original point has already been calculated: it is passed as argument
' fi0. h is a given non-zero increment. Ideally, this h could be varied
' for different kinds of variables.
' ************************************************************************************

xx = x(j%)
x(j%) = x(j%) + h
fi0h = resid(i%, x(), n%)
x(j%) = xx
deriv = (fi0h - fi0) / h
END FUNCTION
```

Topics in time-independent modelling 191

```
SUB gauss (coef(), rhs(), soln(), n%)
DIM aux(n%, n%)
FOR i% = 1 TO n%
soln(i%) = rhs(i%)
FOR j% = 1 TO n%
aux(i%, j%) = coef(i%, j%)
NEXT j%
NEXT i%
IF n% = 1 THEN GOTO 58
FOR k% = 2 TO n%
IF aux(k% - 1, k% - 1) <> 0# THEN GOTO 57
FOR l% = k% TO n%
IF aux(l%, k% - 1) = 0# THEN GOTO 55
FOR m% = k% - 1 TO n%
x = aux(k% - 1, m%)
aux(k% - 1, m%) = aux(l%, m%)
aux(l%, m%) = x
NEXT m%
x = soln(k% - 1)
soln(k% - 1) = soln(l%)
soln(l%) = x
GOTO 57
55 ' keep looking for non - zero pivot
NEXT l%
56 PRINT "singularity detected by Gauss"
STOP
57 ' successful non-zero pivot search. Equations exchanged if necessary.
FOR i% = k% TO n%
x = aux(i%, k% - 1) / aux(k% - 1, k% - 1)
soln(i%) = soln(i%) - x * soln(k% - 1)
FOR j% = k% TO n%
aux(i%, j%) = aux(i%, j%) - x * aux(k% - 1, j%)
NEXT j%
NEXT i%
NEXT k%
58 IF aux(n%, n%) = 0# THEN GOTO 56
soln(n%) = soln(n%) / aux(n%, n%)
IF n% = 1 THEN GOTO 59
FOR k% = n% - 1 TO 1 STEP -1
FOR j% = k% + 1 TO n%
soln(k%) = soln(k%) - aux(k%, j%) * soln(j%)
NEXT j%
soln(k%) = soln(k%) / aux(k%, k%)
NEXT k%
59 ' Solution completed
END SUB

SUB newrap (max%, eps, h, n%, x0())
' ********************************************************************************
```

Appendix A

```
' Newton-Raphson nonlinear solver
' max% = maximum number of iterations
' eps = tolerance
' h = increment for (possibly) numerical differentiation (see SUB deriv)
' n% = number of equations and unknowns
' x0() = initial and updated solution
' deltax() = Newton-Raphson-generated increments of x0()
' *******************************************************************************
DIM coef(n%, n%), rhs(n%), deltax(n%)
FOR k% = 1 TO max%
' Prepare right-hand side vector rhs() and coefficient matrix coef()
FOR i% = 1 TO n%
rhs(i%) = -resid(i%, x0(), n%)
FOR j% = 1 TO n%
coef(i%, j%) = deriv(i%, j%, -rhs(i%), x0(), n%, h)
NEXT j%
NEXT i%
' Solve system of linear equations
CALL gauss(coef(), rhs(), deltax(), n%)
' Update solution and calculate quadratic error
sum1 = 0#
sum2 = 0#
FOR i% = 1 TO n%
x0(i%) = x0(i%) + deltax(i%)
sum1 = sum1 + x0(i%) * x0(i%)
sum2 = sum2 + deltax(i%) * deltax(i%)
NEXT i%
IF (SQR(sum2 / sum1) < = eps) THEN GOTO 99
NEXT k%
PRINT "Convergence not achieved in Newrap after "; max%; " iterations"
STOP
99 ' Convergence achieved
END SUB


FUNCTION resid (i%, x(), n%)
'   Available through COMMON SHARED:
'   np% = number of panels
'   mc% = number of nodes
'   xc(j%,k%) = initial kth coord of jth node
'   forcec = applied force (if any)
'   ynf$ = "y" if right end is fixed, else = "n"
'   fibre() = fibre areas
'   bot() = lower aponeurosis areas
'   top() = upper aponeuroses areas
'   tendon = tendon area
'   activ() = fibre activation coefficients
'   Support conditions enforced directly through equations u = 0.
'   Only panel constraints are treated by means of Lagrange multipliers.
```

Topics in time-independent modelling 193

```
'NOTE: For this version of RESID, MAIN should specify n%= 5*(mc%-1)/2 + 1
IF i%>2 GOTO 11
sum = x(i%)
GOTO 49
11 IF i% > = 2 * (mc% - 1) GOTO 44
sum = 0#
'find left-panel number
pn% = INT((i% - 1) / 4)
pnl% = pn%
pnr% = pn% + 1
IF pnl% = 0 THEN pnl% = pnr%
IF pnl% = np% THEN pnr% = pnl%
'influence of panels on node (derivatives of constraints times multiplier)
FOR jp% = pnl% TO pnr%
v1x = xc(2 * jp% + 2, 1) - xc(2 * jp% - 1, 1)
v1y = xc(2 * jp% + 2, 2) - xc(2 * jp% - 1, 2)
v2x = xc(2 * jp%, 1) - xc(2 * jp% + 1, 1)
v2y = xc(2 * jp%, 2) - xc(2 * jp% + 1, 2)
v1x = v1x + x(4 * jp% + 3) - x(4 * jp% - 3)
v1y = v1y + x(4 * jp% + 4) - x(4 * jp% - 2)
v2x = v2x + x(4 * jp% - 1) - x(4 * jp% + 1)
v2y = v2y + x(4 * jp%) - x(4 * jp% + 2)
ON (i% - 4 * jp% + 4) GOTO 31, 32, 33, 34, 35, 36, 37, 38
31 sum = sum + .5 * (-v2y) * x(2 * mc% + jp%)
GOTO 39
32 sum = sum + .5 * (v2x) * x(2 * mc% + jp%)
GOTO 39
33 sum = sum + .5 * (-v1y) * x(2 * mc% + jp%)
GOTO 39
34 sum = sum + .5 * (v1x) * x(2 * mc% + jp%)
GOTO 39
35 sum = sum + .5 * (v1y) * x(2 * mc% + jp%)
GOTO 39
36 sum = sum + .5 * (-v1x) * x(2 * mc% + jp%)
GOTO 39
37 sum = sum + .5 * (v2y) * x(2 * mc% + jp%)
GOTO 39
38 sum = sum + .5 * (-v2x) * x(2 * mc% + jp%)
39 '
NEXT jp%
'equilibrium influences
'find node number
nn% = INT((i% + 1) / 2)
'find parity
k% = i% - 2 * INT((i% - 1) / 2)
FOR jn% = nn% - 2 TO nn% + 2 STEP 2
IF jn% <= 0 OR jn% >= mc% THEN GOTO 43
jnn% = jn%
```

```
IF jn% <> nn% GOTO 41
jnn% = jn% + 1
IF nn% = 2 * INT(nn% / 2) THEN jnn% = jn% - 1
41 aux = 0#
oldl = 0#
newl = 0#
FOR ks% = 1 TO 2
dij0 = xc(jnn%, ks%) - xc(nn%, ks%)
dij = dij0 + x(2 * jnn% - 2 + ks%) - x(2 * nn% - 2 + ks%)
oldl = oldl + dij0 * dij0
newl = newl + dij * dij
IF ks% = k% THEN aux = aux + dij
NEXT ks%
oldl = SQR(oldl)
newl = SQR(newl)
sum = sum - cons(nn%, jnn%, oldl, newl) * aux / newl
43 NEXT jn%
IF i% <> 2 * mc% - 3 THEN GOTO 49
oldl = xc(mc%, 1) - xc(mc% - 1, 1)
newl = oldl + x(2 * mc% - 1) - x(2 * mc% - 3)
oldl = ABS(oldl)
newl = ABS(newl)
sum = sum - cons(mc% - 1, mc%, oldl, newl)
GOTO 49
44 IF i%>2 * mc% GOTO 46
sum = x(i%)
IF i% = 2 * mc% - 1 THEN sum = sum - forcec
IF ynf$ <> "n" OR i% <> 2 * mc% - 1 THEN GOTO 49
oldl = xc(mc%, 1) - xc(mc% - 1, 1)
newl = oldl + x(2 * mc% - 1) - x(2 * mc% - 3)
oldl = ABS(oldl)
newl = ABS(newl)
sum = cons(mc% - 1, mc%, oldl, newl) - forcec
GOTO 49
46 ' panel constraint equations
' calculate areas as cross-product of panel diagonals
'find panel number
pn% = i% - 2 * mc%
v1x = xc(2 * pn% + 2, 1) - xc(2 * pn% - 1, 1)
v1y = xc(2 * pn% + 2, 2) - xc(2 * pn% - 1, 2)
v2x = xc(2 * pn%, 1) - xc(2 * pn% + 1, 1)
v2y = xc(2 * pn%, 2) - xc(2 * pn% + 1, 2)
area = .5 * (v1x * v2y - v1y * v2x)
v1x = v1x + x(4 * pn% + 3) - x(4 * pn% -3)
v1y = v1y + x(4 * pn% + 4) - x(4 * pn% -2)
v2x = v2x + x(4 * pn% - 1) - x(4 * pn%  + 1)
v2y = v2y + x(4 * pn%) - x(4 * pn% + 2)
area1 = .5 * (v1x * v2y - v1y * v2x)
```

```
         sum = areal - area
49       resid = sum
         END FUNCTION
```

A.8 EXAMPLE: STATIC DEFORMATION OF A CAT MEDIAL GASTROCNEMIUS MUSCLE

This problem has already been formulated in Chapter 6. What follows is a verbatim reproduction of an interactive session with the program as well as a listing of the resulting output file.

```
This is a program for the static solution of a straight-line planar
muscle model. Follow the self-explanatory input instructions.

                GENERAL APPEARANCE


y-axis                       o:node
 ^       2                   f:fibre
 |      o    4               a:aponeurosis
 |     /   a   o
 |    /     / a   6
 |   f/    f/    o
 |   / panel /   f/   .
 |  /  # 1 / panel /    . 2p                 2p + 3
 | /      /   #2  /    o    2p+2           (roller & force
 |/      /      /    f/   a (roller)        or fixed)
 o-------/--------/----------------/---------o------------------o------- >
 1   a   o      /            / panel /       tendon           x-axis
(fixed)  3    a o            /  # p /f
              5    '       /       /
                         o    a    o
                        2p-1      2p+1


enter number of panels : 3

next, you will enter the nodal coordinates in the x-y system depicted
           (in mm)
for node #    1    x = 0     y = 0
for node #    2    x = 18.1  y = 8.5
for node #    3    x = 19.7  y = -3.5
for node #    4    x = 41    y = 5.1
for node #    5    x = 39.4  y = -6.9
for node #    6    x = 64.1  y = 1.1
for node #    7    x = 59    y = -10.4
for node #    8    x = 88    y = 0
for node #    9    x = 120   y = 0
is the last node fixed, i.e., displacement controlled (y),
or is it movable, i.e., force controlled (n)? Enter y or n: n
```

196 Appendix A

next you will enter the fibre-, aponeuroses- and tendon-areas of influence
                    (in sq mm)
fibre 1 - 2 : 75 apo. 1 - 3 : 2 apo. 2 - 4 : 4
fibre 3 - 4 : 150 apo. 3 - 5 : 3 apo. 4 - 6 : 3
fibre 5 - 6 : 150 apo. 5 - 7 : 4 apo. 6 - 8 : 2
fibre 7 - 8 : 75 tendon: 4
next, you will enter the ratio: initial length to optimal length
(This ratio is assumed to be the same for all fibres) lr0 = 0.8
next, you will enter parameters to control the numerical procedure:
  max = maximum number of iterations in Newton-Raphson (say, 50)
  eps = relative quadratic error for convergence (say, 0.005)
  h   = increment for numerical differentiation (say, 0.001)
max = 50 eps = 0.002 h = 0.001
Finally, you will enter the loading conditions
You have specified a movable end. Therefore, enter the force (in N):10
You will now enter the fibre activation coefficients (between 0 and 1)
fibre 1 - 2 : 0.1
fibre 3 - 4 : 0.1
fibre 5 - 6 : 0.1
fibre 7 - 8 : 0.1
 continue? y or n
? y
You have decided to continue. Present values will be used as initial
You will now enter the new loading. Follow instructions as before
You have specified a movable end. Therefore, enter the force (in N):20
You will now enter the fibre activation coefficients (between 0 and 1)
fibre 1 - 2 : 0.2
fibre 3 - 4 : 0.2
fibre 5 - 6 : 0.2
fibre 7 - 8 : 0.2
 continue? y or n
? y
You have decided to continue. Present values will be used as initial
You will now enter the new loading. Follow instructions as before
You have specified a movable end. Therefore, enter the force (in N):50
You will now enter the fibre activation coefficients (between 0 and 1)
fibre 1 - 2 : 0.6
fibre 3 - 4 : 0.6
fibre 5 - 6 : 0.6
fibre 7 - 8 : 0.6
 continue? y or n
? y
You have decided to continue. Present values will be used as initial
You will now enter the new loading. Follow instructions as before
You have specified a movable end. Therefore, enter the force (in N):100
You will now enter the fibre activation coefficients (between 0 and 1)
fibre 1 - 2 : 1.
fibre 3 - 4 : 1.

Topics in time-independent modelling

```
fibre 5 - 6 : 1.
fibre 7 - 8 : 1.
 continue? y or n
? n
solution for applied force 10

FIBRE SPECIFIED ACTIVATION COEFFICIENT
1 2 .1
3 4 .1
5 6 .1
7 8 .1


NODE      X-DISPL          Y-DISPL
1      0          0
2      .4617181471102775    1.104899654488119
3      .5115359409572214    1.30777478133989
4      .5550129827399424    .9190421890535039
5      .6684374436518781    1.018573559140207
6      .7747373418630168    .7930210718113649
7      .7649164965448635    1.001097062198351
8      1.109064585769029    0
9      1.497452887794715    0

FIBRE FORCE APO FORCE APO FORCE
1 2 2.115260375154882  1 3 8.538632734628004  2 4 2.091177388185914
3 4 3.655244977749466  3 5 5.481316351394956  4 6 5.363611635959358
5 6 3.721453357942368  5 7 1.903603654465359  6 8 8.467221541294094
7 8 1.854189434965889 calculated tendon force =   10.00005156514882

solution for applied force 20

FIBRE SPECIFIED ACTIVATION COEFFICIENT
1 2 .2
3 4 .2
5 6 .2
7 8 .2


NODE X-DISPL Y-DISPL
1      0          0
2      .5018098837057897    1.058713298751827
3      .6506710819140974    1.304528261064555
4      .6575868835983332    .9090621960081814
5      .904430869954665     1.039910400706473
6      .9872267084061495    .8023325215772592
7      1.053709483684407    1.054147375520221
8      1.486308119206001    0
9      2.04472347679578     0
```

FIBRE   FORCE   APO   FORCE   APO   FORCE
1 2 4.238909874895694 1 3 17.05481401521497 2 4 4.190613207172147
3 4 7.273595479323157 3 5 10.96365961553815 4 6 10.70421285906729
5 6 7.419658548738086 5 7 3.831834864255694 6 8 16.90323482705982
7 8 3.734031727426871 calculated tendon force = 20.00041033385446

solution for applied force 50

FIBRE SPECIFIED ACTIVATION COEFFICIENT
1 2 .6
3 4 .6
5 6 .6
7 8 .6

NODE X-DISPL Y-DISPL
1   0   0
2   -.7882891257758402   1.084998916063293
3   .9227640070088361   1.242070946778246
4   -.4961946796024357   1.084241172568661
5   1.375251938279308   1.035210845066135
6   3.456778592058384D-02   .9337312080461689
7   1.635626211265825   1.112874468083716
8   .848517690288112   0
9   1.744483587919631   0

FIBRE   FORCE   APO   FORCE   APO   FORCE
1 2 10.75776232026041 1 3 42.81409523138088 2 4 10.55251602323538
3 4 18.08660829606774 3 5 27.96855650023027 4 6 26.45975629951866
5 6 18.74426851960263 5 7 10.10899882730048 6 8 41.89605108551913
7 8 9.871128242083783 calculated tendon force = 50.00000864813935

solution for applied force 100

FIBRE   SPECIFIED   ACTIVATION   COEFFICIENT
1 2 1
3 4 1
5 6 1
7 8 1

NODE   X-DISPL   Y-DISPL
1   0   0
2   .6708555173475761   .8668246162373286
3   1.236031064394385   1.290914645398239
4   1.091028845770359   .8662780757355473
5   1.897869679490133   1.128641402149792
6   1.881831778240251   .8404375811349497
7   2.272445182116604   1.272923795466991
8   3.073236278672614   0
9   4.349732681872501   0

Topics in time-independent modelling

```
FIBRE  FORCE              APO  FORCE              APO  FORCE
1  2  21.3804761200064    1  3  84.80198510001489    2  4  21.13779414580502
3  4  35.59940682692327   3  5  54.84021602428393    4  6  53.04975375789297
5  6  36.60946215972508   5  7  19.67029303762074    6  8  83.85873774234501
7  8  19.20910556046774   calculated tendon force =  100.0010518525333
```

A.9 A VARIANT OF THE PREVIOUS EXAMPLE

The following are the results obtained for the same geometric and activation data as before, but keeping the external force fixed at 10 N.

```
solution for applied force 10

FIBRE   SPECIFIED   ACTIVATION   COEFFICIENT
1 2 .1
3 4 .1
5 6 .1
7 8 .1


NODE     X-DISPL           Y-DISPL
1        0                 0
2        .4617182089900064  1.104899614072247
3        .5115359432742994  1.307774762662603
4        .555013054668743   .9190422299477098
5        .6684374646787394  1.018573594834253
6        .7747373891184902  .7930210559313547
7        .7649164979018266  1.001097028394595
8        1.109064660268218  0
9        1.497452962293606  0


FIBRE  FORCE              APO  FORCE              APO  FORCE
1  2  2.115260379300845   1  3  8.53863271219225    2  4  2.091177400442558
3  4  3.655245018091164   3  5  5.481316353879534   4  6  5.363611703923276
5  6  3.721453359429991   5  7  1.9036036772475     6  8  8.467221509743716
7  8  1.854189452355685   calculated tendon force =  10.00005156513427


solution for applied force 10

FIBRE SPECIFIED ACTIVATION COEFFICIENT
1 2 .2
3 4 .2
5 6 .2
7 8 .2


NODE     X-DISPL           Y-DISPL
1        0                 0
2        -3.849505343050651  1.489328464520286
3        .4979560826145958   1.14632020805176
```

```
4    -3.710358004719851    1.607675412129704
5     .6730510782348491     .9106133807175346
6    3.542704660902291    1.215043674162162
7     .7776651237949646    .8887475181482786
8    3.248589790617213    0
9   -2.86020244114707     0


FIBRE   FORCE   APO   FORCE   APO   FIBRE


1 2 2.20270346803201  1 3 8.772053546720022  2 4 2.105570935638557
3 4 3.75290146671736  3 5 5.899109418304604  4 6 5.226568060201583
5 6 3.96566028162403  5 7 2.217556567410417  6 8 8.318293676657362
7 8 2.169207257896615 calculated tendon force = 10.00000513051077

solution for applied force 10

FIBRE    SPECIFIED    ACTIVATION    COEFFICIENT

1 2 .6
3 4 .6
5 6 .6
7 8 .6


NODE    X-DISPL       Y-DISPL

1    0            0
2   -7.269095593107411    1.772332339619281
3    .4971256224379579    1.088608521379414
4   -7.069449522144887    2.351699457121186
5    .7178172359666745    1.051813673481401
6   -6.917360150638109    1.90892381402086
7    .8437992612941992    1.06784051555018
8   -6.706947830320662    0
9   -6.318538249610576    0


FIBRE   FORCE   APO   FORCE   APO   FORCE
1 2 2.301175521279527  1 3 9.081009949970234  2 4 2.090935779931232
3 4 3.876990796591407  3 5 6.434494537556046  4 6 4.982098201004731
5 6 4.186387394261171  5 7 2.783921984912717  6 8 8.019264856353136
7 8 2.744765207491169 calculated tendon force =  10.00108887440611

solution for applied force 10

FIBRE    SPECIFIED    ACTIVATION    COEFFICIENT
1 2 1
3 4 1
5 6 1
7 8 1
```

```
NODE    X-DISPL          Y-DISPL
1       0                0
2       -8.072647433074117   1.840452415915163
3       .5011552064565886    1.106454639253563
4       -7.855246925186843   2.576066726714443
5       .741222554183347     1.164344510080268
6       -7.692847138847982   2.202930321745824
7       .8787196869882912    1.212365830405152
8       -7.524468411034803   0
9       -7.136080355137953   0


FIBRE  FORCE   APO   FORCE   APO   FORCE
1 2 2.317140650950775 1 3 9.180690232937714 2 4 2.066723136184973
3 4 3.891912218648995 3 5 6.610642083803983 4 6 4.867248139148499
5 6 4.217170701906682 5 7 3.029945655283574 6 8 7.876063891450209
7 8 2.996733542066481 calculated tendon force =    10.00003956698963
```

Appendix B
Topics in time-dependent modelling

B.1 THE FINITE-DIFFERENCE METHOD

In this section we will deal with the most common and versatile discretization technique in the time domain, namely, the finite-difference method. Starting from an initial time t_0, for which the **initial conditions** of the system are known, the future is partitioned into intervals of equal[1] length h and the solution is sought for the discrete times $t_0 + h$, $t_0 + 2h$, $t_0 + 3h$, ..., sometimes referred to as the **mesh points**. A large variety of finite-difference techniques exist, but they can be classified into two main categories: explicit and implicit. To explain the essential features of these two categories, let us assume that our dynamic equations have the form:

$$f_I\left(\mathbf{x}, \frac{d\mathbf{x}}{dt}\right) = m_I \frac{d^2 x_I}{dt^2}, \quad I = 1,...,n \quad (B.1)$$

where f_I are given functions and $\mathbf{x}$ represents the vector of independent variables $x_1, ..., x_n$, all of which are assumed to be functions of time t, and where m_I is a positive mass associated with the Ith degree of freedom.[2] In an **explicit** method, from the (approximate) knowledge of the solution $\mathbf{x}$ at times $t, t-h, t-2h, ...$, the approximate value of $\mathbf{x}$ at time $t + h$ is obtained by enforcing (approximately) the validity of equation (B.1) at time t. In an **implicit** method, on the other hand, the (approximate) value of $\mathbf{x}$ at time $t + h$ is obtained by enforcing this equation at time $t + h$.

The simplest explicit scheme makes use of **central differences**, whereby the first and second derivatives at time t are approximated, respectively, by the formulae

$$\left[\frac{d\mathbf{x}}{dt}\right]_t \simeq \frac{\mathbf{x}_{t+h} - \mathbf{x}_{t-h}}{2h} \quad (B.2)$$

---
[1] Variants of the method for unequal time intervals can be formulated without much extra effort.
[2] This is not the most general form that dynamic equations may have, but is general enough for the purposes of describing the method.

and
$$\left[\frac{d^2 \mathbf{x}}{dt^2}\right]_t \cong \frac{\mathbf{x}_{t+h} - 2\mathbf{x}_t + \mathbf{x}_{t-h}}{h^2} \qquad (B.3)$$

where the subscript indicates the time at which the function is evaluated. These formulae correspond to assuming a parabolic variation of $\mathbf{x}$ in the time interval $(t - h, t + h)$. Substituting into equation (B.1), we obtain

$$f_I\left(\mathbf{x}_t, \frac{\mathbf{x}_{t+h} - \mathbf{x}_{t-h}}{2h}\right) = m_I \frac{x_{I,t+h} - 2x_{I,t} + x_{I,t-h}}{h^2}, \qquad i = 1,\ldots,n \qquad (B.4)$$

a system of n algebraic equations for the n unknowns represented by the vector $\mathbf{x}_{t+h}$. To get the procedure started, we note that at time t_0 both $\mathbf{x}$ and $d\mathbf{x}/dt$ are known as initial conditions, but equation (B.3) would require a value of $\mathbf{x}$ at time $t_0 - h$. A fictitious value for this $\mathbf{x}_{t_0 - h}$ can be found consistent with the initial conditions and the differential equations by noting that at the initial time equation (B.1) delivers

$$\left[\frac{d^2 x_I}{dt^2}\right]_{t_0} = \frac{f_I\left(\mathbf{x}_{t_0}, \left[\frac{d\mathbf{x}}{dt}\right]_{t_0}\right)}{m_I} \qquad (B.5)$$

a value that, when used in conjunction with equations (B.2) and (B.3), yields

$$\mathbf{x}_{t_0-h} = \mathbf{x}_{t_0} - h\left[\frac{d\mathbf{x}}{dt}\right]_{t_0} + \frac{1}{2} h^2 \left[\frac{d^2 x_I}{dt^2}\right]_{t_0} \qquad (B.6)$$

An interesting feature of the explicit method can be appreciated considering the particular case in which the differential equations happen not to contain a velocity dependence. Then, the solution of the algebraic equations (B.4) is simply

$$\mathbf{x}_{t+h} = 2\mathbf{x}_t - \mathbf{x}_{t-h} + h^2 \frac{f_I(\mathbf{x}_t)}{m_I} \qquad (B.7)$$

We thus face the seemingly paradoxical situation that dynamics is simpler than statics! This is indeed the case for the system considered, but we should bear in mind that statics does not become, in the framework of this explicit procedure, a particular case of dynamics, in the sense that it does not emerge in the approximate solution (as it does in the exact differential equations) as a particular case of dynamics when all masses vanish. The approximate solution (B.7) would, rather, become infinitely large. This fact also hints at the phenomenon of the so-called **conditional stability** of the method: for the procedure to be numerically stable the time increment h must be smaller than a certain value which depends directly on the mass-to-stiffness ratio. In spite of this limitation, the explicit method of central differences is a very useful technique, characterized by the

Appendix B

simplicity of its step-by-step solution. Convergence may be judged by refining the mesh, i.e. the size of the step: if the approximate solutions corresponding to step sizes, say, h and $h/2$ differ very little on their common time points, then the solution may be deemed satisfactory. To illustrate some of these concepts, we investigate the following one-degree-of-freedom example.

B.2 EXAMPLE: DYNAMICS OF A ONE-DEGREE-OF-FREEDOM SYSTEM BY CENTRAL DIFFERENCES

Let us solve numerically the problem of free vibrations of a linear spring-and-mass system (Figure B.1) by the explicit method of central differences. We shall use several time increments and comment on numerical stability.

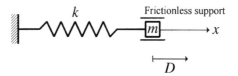

Figure B.1 Flow chart.

In the absence of applied external forces, the differential equation of motion is:

$$m\frac{d^2x}{dt^2} + kx = 0 \tag{B.8}$$

where x is measured from the static equilibrium position, k is the spring constant, and m is the mass. For definiteness, we assume the initial conditions at $t = 0$ to be

$$x_0 = D \ne 0, \qquad \left[\frac{dx}{dt}\right]_0 = 0 \tag{B.9}$$

The exact solution of (B.8) with the initial conditions (B.9) is:

$$x = D \cos \omega t \tag{B.10}$$

where

$$\omega = \left(\frac{k}{m}\right)^{1/2} \tag{B.11}$$

is the **natural frequency**, governed by the ratio between the elasticity (k) and the inertia (m). Exact solutions directly available as in this case are rare. To seek an approximate solution we start by plugging the initial conditions (B.9) into the differential equation (B.8), to obtain:

$$\left[\frac{d^2x}{dt^2}\right]_0 = -D\omega^2 \tag{B.12}$$

whence, by expression (B.6):

$$x_{-h} = D(1 - 0.5h^2\omega^2) \qquad (B.13)$$

Using now the general formula (B.7), we obtain successively the values for x_h, x_{2h}, x_{3h}, ..., etc. It is apparent that whatever features this numerical algorithm might exhibit will be governed by the dimensionless parameter $\alpha = h\omega$. It is not difficult to attach to α a physical meaning: we have already identified ω as the natural frequency of the system. The **natural period**, i.e. the time needed for one full oscillation to take place, is then given by:

$$T = \frac{2\pi}{\omega} \qquad (B.14)$$

Therefore, the parameter α governing the numerical procedure can be written as

$$\alpha = \frac{2\pi h}{T} \qquad (B.15)$$

showing that α is a measure of the ratio between the chosen time increment h and the natural time scale T of the oscillation, governed in its turn by the inertia-to-elasticity ratio of the system. There are now two questions to be addressed. First, is the algorithm numerically stable? Second, if so, does its solution mimic accurately the exact solution of the differential equation? The answer to both of these questions depends on the magnitude of α. The meaning of **numerical stability** in its general sense is roughly as follows: the algorithmn is stable if a small numerical perturbation tends to be attenuated with increasing time. If, on the contrary, the perturbation grows without bound as time goes on, we say that the scheme has become numerically unstable. In a one-degree-of-freedom system, instability results in the solution itself growing without bound. If, as in our case, stability turns out to depend on the size of the time step, we say that the scheme is **conditionally stable**. It can be shown by theoretical means that in the system at hand stability is guaranteed for $\alpha < 2$. We emphasize, however, that stability does not in itself ensure faithfulness to the exact solution: it merely asserts the insensitivity of the numerical scheme to round-off errors. For faithfulness we will require, in general, even smaller values of α, as can be intuitively understood when realizing that $\alpha = 2$ would only sample about three points per full oscillation. To get a better grasp of these concepts, Table B.1 compares the results for various choices of α (i.e. of h), extending the solution over two complete periods. We have listed the exact solution (to four digits), for $D = 1$, every $0.5/\omega$ seconds, and the chosen values of α are 0.25, 0.5, 1.0, 1.98, 2.02 and 4.0. Naturally, for $\alpha = 0.25$ the numerical solution produces twice as many values than those shown, but these intermediate values have not been listed.

We notice that the solution for $\alpha = 0.25$ is very accurate during the entire first two periods, while for $\alpha = 0.5$ the accuracy is quite good during the first period only. If we were to plot these results against the exact solution, we would observe that the inaccuracy is translated basically into a slight and progressive

Table B.1 Method of central differences: comparative results for different values of α

Exact	α = 0.25	α = 0.50	α = 1.00	α = 1.98	α = 2.02	α = 4.00
0.8776	0.8770	0.8750				
0.5403	0.5381	0.5313	0.5			
0.0707	0.0067	0.0547				
−0.4151	−0.4209	−0.4355	−0.5	0.9602	−1.040	
−0.8011	−0.8051	−0.8169				
−0.9900	−0.9911	−0.9940	−1			
−0.9365	−0.9332	−0.9226				
−0.6536	−0.6457	−0.6206	−0.5	0.8440	1.164	−7
−0.2108	−0.1992	−0.1634				
0.2837	0.2962	0.3346	0.5			
0.7087	0.7188	0.7490				
0.9602	0.9644	0.9761	1	−0.6606	−1.381	
0.9766	0.9728	0.9591				
0.7539	0.7417	0.7025	0.5			
0.3466	0.3281	0.2702				
−0.1455	−0.1662	−0.2297	−0.5	0.4246	1.710	97
−0.6020	−0.6197	−0.6722				
−0.9111	−0.9206	−0.9466	−1			
−0.9972	−0.9950	−0.9844				
−0.8391	−0.8245	−0.7760	−0.5	−0.1548	−2.176	
−0.4755	−0.4511	−0.3737				
0.0044	0.0033	0.1221	0.5			
0.4833	0.5095	0.5873				
0.8439	0.8603	0.9057	1	−0.1273	2.817	−1351
0.9978	0.9994	0.9977				

contraction of the period. For α = 1.0, the solution is bounded but inaccurate. For α = 1.98, we observe that the solution is still bounded, as guaranteed by the theoretical criterion of conditional stability, but completely wrong, while for α slightly larger than the limiting value of 2 the solution grows without bound.

B.3 TIME-DEPENDENT PROBLEMS: IMPLICIT METHODS

From the preceding example, it can be understood that for many problems in dynamics it may be desirable to eliminate the two main disadvantages of explicit methods, namely, conditional stability and the fact that statics is not somehow recoverable from dynamics (when letting the masses become very small). Implicit procedures provide, at least in part, an answer to both drawbacks. In many implicit schemes it is assumed that during an interval of time starting at the generic mesh point $t_i = t_0 + ih$ the variation of the displacements **x** is a cubic polynomial. This is tantamount to assuming that the acceleration varies linearly

Topics in time-dependent modelling

within that interval. A generic cubic polynomial abides by the formula:

$$\mathbf{x} = \mathbf{a} + \mathbf{b}(t - t_i) + \mathbf{c}(t - t_i)^2 + \mathbf{d}(t - t_i)^3 \tag{B.16}$$

where **a**, **b**, **c**, **d** are vectors comprised of constants. The various implicit techniques differ in the way these constants are controlled. Here we will only present two such techniques – Houbolt's method and Wilson's θ-method.

In **Houbolt's method** [B.1] the cubic polynomial is assumed to fit the known present time t_i, as well as the two immediately preceding mesh points t_{i-1} and t_{i-2}. The extra condition for the polynomial to pass through the next mesh point, t_{i+1}, is obtained by enforcing (approximately) the equations of motion (B.1) at time $t_{i+1} = t_i + h$. The interpolation conditions for the cubic (B.16) to pass through the aforementioned four points are:

$$\mathbf{x}_{i-2} = \mathbf{a} + \mathbf{b}(-2h) + \mathbf{c}(-2h)^2 + \mathbf{d}(-2h)^3 \tag{B.17}$$

$$\mathbf{x}_{i-1} = \mathbf{a} + \mathbf{b}(-h) + \mathbf{c}(-h)^2 + \mathbf{d}(-h)^3 \tag{B.18}$$

$$\mathbf{x}_i = \mathbf{a} \tag{B.19}$$

$$\mathbf{x}_{i+1} = \mathbf{a} + \mathbf{b}(h) + \mathbf{c}(h)^2 + \mathbf{d}(h)^3 \tag{B.20}$$

yielding, by direct linear elimination,

$$\mathbf{a} = \mathbf{x}_i \tag{B.21}$$

$$\mathbf{b} = \frac{\mathbf{x}_{i-2} - 6\mathbf{x}_{i-1} + 3\mathbf{x}_i + 2\mathbf{x}_{i+1}}{6h} \tag{B.22}$$

$$\mathbf{c} = \frac{\mathbf{x}_{i-1} - 2\mathbf{x}_i + \mathbf{x}_{i+1}}{2h^2} \tag{B.23}$$

$$\mathbf{d} = \frac{-\mathbf{x}_{i-2} - 3\mathbf{x}_{i-1} - 3\mathbf{x}_i + \mathbf{x}_{i+1}}{6h^3} \tag{B.24}$$

Differentiating equation (B.16) and then setting $t = t_i + h$, we obtain the following expressions for the first and second derivatives of **x** at time t_{i+1}:

$$\left[\frac{d\mathbf{x}}{dt}\right]_{i+1} = \mathbf{b} + 2\mathbf{c}h + 3\mathbf{d}h^2 = \frac{-2\mathbf{x}_{i-2} + 9\mathbf{x}_{i-1} - 18\mathbf{x}_i + 11\mathbf{x}_{i+1}}{6h} \tag{B.25}$$

and

$$\left[\frac{d^2\mathbf{x}}{dt^2}\right]_{i+1} = 2\mathbf{c} + 6\mathbf{d}h = \frac{-\mathbf{x}_{i-2} + 4\mathbf{x}_{i-1} - 5\mathbf{x}_i + 2\mathbf{x}_{i+1}}{h^2} \tag{B.26}$$

Upon introducing these formulae in the equations of motion (B.1), a system of algebraic equations for the vector $\mathbf{x}_{i+1}$ is obtained:

$$f_I\left[\mathbf{x}_{i-1}, \frac{-2\mathbf{x}_{i-2} + 9\mathbf{x}_{i-1} - 18\mathbf{x}_i + 11\mathbf{x}_{i+1}}{6h}\right] = m_I \frac{x_{I,i-2} + 4x_{I,i-1} - 5x_{I,i} + 2x_{I,i+1}}{h^2} \tag{B.27}$$

where $x_{I,i}$ denotes the Ith entry of the vector $\mathbf{x}_i$. Notice that, unlike its counterpart (B.4) in the explicit method, this system can be regarded as a modification of the static equations. Indeed, if we consider, for example, the case in which there is no velocity dependence, the resulting equations are a direct restatement of the equations of statics with an extra 'force' given by the term $m_I(-x_{I,i-2} + 4x_{I,i-1} - 5x_{I,i})/h^2$, and a 'stiffness' increase by the amount $2m_I/h^2$. Both terms vanish if the mass does and, moreover, the scheme can be shown to be unconditionally stable (for all values of h, no matter how large). To start up the procedure, however, we need to give special consideration to the first two time steps. Let $\mathbf{x}_0$ and $[d\mathbf{x}/dt]_0$ be the known initial displacements and velocities. If the masses are not zero, equation (B.1) yields also the initial accelerations $[d^2\mathbf{x}/dt^2]_0$, just as in the explicit method. Differentiating the basic polynomial (B.16) at $t = t_i = t_0$, we arrive at:

$$\left[\frac{d\mathbf{x}}{dt}\right]_0 = \mathbf{b} = \frac{\mathbf{x}_{-2} - 6\mathbf{x}_{-1} + 3\mathbf{x}_0 + 2\mathbf{x}_1}{6h} \tag{B.28}$$

and

$$\left[\frac{d^2\mathbf{x}}{dt^2}\right]_0 = 2\mathbf{c} = \frac{\mathbf{x}_{-1} - 2\mathbf{x}_0 + \mathbf{x}_1}{h^2} \tag{B.29}$$

whence, by elimination,

$$\mathbf{x}_{-1} = h^2 \left[\frac{d^2\mathbf{x}}{dt^2}\right]_0 + 2\mathbf{x}_0 - \mathbf{x}_1 \tag{B.30}$$

and

$$\mathbf{x}_{-2} = 6h^2 \left[\frac{d^2\mathbf{x}}{dt^2}\right]_0 + 6h\left[\frac{d\mathbf{x}}{dt}\right]_0 + 9\mathbf{x}_0 - 8\mathbf{x}_1 \tag{B.31}$$

Finally, using these formulae in conjunction with equations (B.25) and (B.26) for $i = 0$ and $i = 1$, yields the following special formulae for the first two steps:

$$\left[\frac{d\mathbf{x}}{dt}\right]_1 = \frac{-h^2\left[\frac{d^2\mathbf{x}}{dt^2}\right]_0 - 4h\left[\frac{d\mathbf{x}}{dt}\right]_0 - 6\mathbf{x}_0 + 6\mathbf{x}_1}{2h} \tag{B.32}$$

$$\left[\frac{d^2\mathbf{x}}{dt^2}\right]_1 = \frac{-2h^2\left[\frac{d^2\mathbf{x}}{dt^2}\right]_0 - 6h\left[\frac{d\mathbf{x}}{dt}\right]_0 - 6\mathbf{x}_0 + 6\mathbf{x}_1}{h^2} \tag{B.33}$$

$$\left[\frac{d\mathbf{x}}{dt}\right]_2 = \frac{-2h^2\left[\frac{d^2\mathbf{x}}{dt^2}\right]_0 + 5\mathbf{x}_0 - 16\mathbf{x}_1 + 11\mathbf{x}_2}{6h} \tag{B.34}$$

Topics in time-dependent modelling

$$\left[\frac{d^2\mathbf{x}}{dt^2}\right]_2 = \frac{-h^2\left[\frac{d^2\mathbf{x}}{dt^2}\right]_0 + 2\mathbf{x}_0 - 4\mathbf{x}_1 + 2\mathbf{x}_2}{h^2} \quad (B.35)$$

Other starting procedures can be advocated, such as the use of an explicit scheme for the first two steps only.

While unconditionally stable, Houbolt's method tends to introduce a relatively large amount of so-called numerical damping, resulting in a progressive amplitude decrease of an otherwise undamped vibration. Wilson's θ-method [2], on the other hand, exhibits a better behaviour in that respect and is, moreover, free of special starting formulae, provided the acceleration at the initial time is read off the differential equations, as we have also done for the other two methods discussed so far. Starting again from the general cubic:

$$\mathbf{x} = \mathbf{a} + \mathbf{b}(t - t_i) + \mathbf{c}(t - t_i)^2 + \mathbf{d}(t - t_i)^3 \quad (B.36)$$

we compute the function and its first and second derivatives at time $t = t_i$ as

$$\mathbf{x}_i = \mathbf{a} \quad (B.37)$$

$$\left[\frac{d\mathbf{x}}{dt}\right]_i = \mathbf{b} \quad (B.38)$$

$$\left[\frac{d^2\mathbf{x}}{dt^2}\right]_i = 2\mathbf{bc} \quad (B.39)$$

so that the cubic polynomial (B.36) and its derivatives at any subsequent time t can be rewritten as:

$$\mathbf{x}(t) = \mathbf{x}_i + \left[\frac{d\mathbf{x}}{dt}\right]_i (t - t_i) + \frac{1}{2}\left[\frac{d^2\mathbf{x}}{dt^2}\right]_i (t - t_i)^2 + \mathbf{d}(t - t_i)^3 \quad (B.40)$$

$$\frac{d\mathbf{x}(t)}{dt} = \left[\frac{d\mathbf{x}}{dt}\right]_i + \left[\frac{d^2\mathbf{x}}{dt^2}\right]_i (t - t_i) + 3\mathbf{d}(t - t_i)^2 \quad (B.41)$$

$$\frac{d^2\mathbf{x}(t)}{dt^2} = \left[\frac{d^2\mathbf{x}}{dt^2}\right]_i + 6\mathbf{d}(t - t_i) \quad (B.42)$$

This implies that, assuming $\mathbf{x}$ and its first and second derivatives to be known at time t_i, we are left with just $\mathbf{d}$ to be determined by enforcing the equations of motion at some later point of time. Instead of $\mathbf{d}$, it is more convenient to use $\mathbf{x}_{i+1}$ as the quantity to be determined. It follows directly from equation (B.40) that

$$\mathbf{x}_{i+1} = \mathbf{x}_i + \left[\frac{d\mathbf{x}}{dt}\right]_i h + \frac{1}{2}\left[\frac{d^2\mathbf{x}}{dt^2}\right]_i h^2 + \mathbf{d}h^3 \quad (B.43)$$

whence

$$\mathbf{d} = \frac{\mathbf{x}_{i+1} - \mathbf{x}_i - \left[\dfrac{d\mathbf{x}}{dt}\right]_i h - \dfrac{1}{2}\left[\dfrac{d^2\mathbf{x}}{dt^2}\right]_i h^2}{h^3} \qquad (B.44)$$

which, when plugged into equations (B.40), (B.41), and (B.42), yields the following expressions for the function and its first two derivatives:

$$\mathbf{x}(t) = \mathbf{x}_i(1 - \theta^3) + \left[\frac{d\mathbf{x}}{dt}\right]_i \theta h(1 - \theta^2) + \frac{1}{2}\left[\frac{d^2\mathbf{x}}{dt^2}\right]_i \theta^2 h^2(1 - \theta) + \mathbf{x}_{i+1} \qquad (B.45)$$

$$\frac{d\mathbf{x}(t)}{dt} = \frac{-3\mathbf{x}_i \theta^2 + \left[\dfrac{d\mathbf{x}}{dt}\right]_i h(1 - 3\theta^2) + \dfrac{1}{2}\left[\dfrac{d^2\mathbf{x}}{dt^2}\right]_i \theta h^2(2 - 3\theta) + 3\mathbf{x}_{i+1}}{h} \qquad (B.46)$$

$$\frac{d^2\mathbf{x}(t)}{dt^2} = \frac{-6\mathbf{x}_i \theta - 6\left[\dfrac{d\mathbf{x}}{dt}\right]_i \theta h + \left[\dfrac{d^2\mathbf{x}}{dt^2}\right]_i h^2(1 - 3\theta) + 6\mathbf{x}_{i+1}\theta}{h^2} \qquad (B.47)$$

where the time variable has been replaced by the new variable

$$\theta = \frac{t - t_i}{h} \qquad (B.48)$$

There only remains to choose a point of time at which to enforce the equations of motion so as to determine the unknown value of $\mathbf{x}_{i+1}$. An obvious choice would be $\theta = 1$, i.e. the next mesh point t_{i+1}, but it can be shown that this would lead to an algorithm which is only conditionally stable.[3] Instead, theoretical considerations show that for values of θ greater than 1.37 the algorithm becomes unconditionally stable. Choosing, for instance, $\theta = 1.40$, and introducing equations (B.46), (B.47), and (B.48) into (B.1), a system of algebraic equations is obtained for $\mathbf{x}_{i+1}$.[4] Solving this system, the new $\mathbf{x}_{i+1}$ is used in equations (B.47) and (B.48) with $\theta = 1$ to produce the values necessary to restart the process for the

---

[3] The reader will notice that this is precisely what was done for the first two steps of Houbolt's method. Indeed, equations (B.46) and (B.47) reproduce exactly (B.32) and (B.33) upon using $i = 0$ and $\theta = 1$. It can then be concluded that the suggested starting procedure may introduce an element of instability in Houbolt's method if h is not small enough.

[4] It is customary to present Wilson's θ-method through an intermediate step whereby $\mathbf{x}$ is first solved at time $t_i + \theta h$ and this value is then used to obtain $\mathbf{x}$ at time $t_{i+1} = t_i + h$ by means of the basic cubic interpolation formula. Our presentation is completely equivalent. The important fact is that the equations of motion themselves be imposed at time $t_i + \theta h$. When solving *linear* dynamic equations, it is also customary to extrapolate linearly the load vector from times t_i and t_{i+1} to $t_i + \theta h$. In the generality of our equation (B.1) this is not quite possible since the 'load vector' has not been separated from the functions f_l. In practice, however, it is usually possible to identify the external load portion of these functions, in which case the linear extrapolation is desirable.

Topics in time-dependent modelling 211

next time step. We see that one of the attractive features of Wilson's θ-method is that, except for the fact that the initial acceleration is to be obtained for the first step from the equations of motion, there is no difference between the first step and the succeeding ones, so that no special starting procedure is actually needed, in contrast with Houbolt's method.

For comparison purposes, Table B.2 presents the numerical results obtained with Wilson's θ-method for the linear spring-and-mass system of the previous example. It can be seen that (with $\theta = 1.4$) the scheme is stable for even very large values of h. The quality of the results, however, in this particular example was not as good as with the central-difference method. To achieve similar quality we had to resort to a value of $\alpha = 0.125$, as shown in the last column of the table, which was added for the sake of fairness in comparative assessment.

The real need for the use of unconditionally stable methods arises in the solution of systems with many degrees of freedom. To understand why this is the case, it is only necessary to remember that a (linear) system has as many natural fre-

Table B.2 Wilson's θ-method: comparative results for different values of α

Exact	$\alpha = 0.25$	$\alpha = 0.50$	$\alpha = 1.00$	$\alpha = 2.00$	$\alpha = 4.00$	$\alpha = 0.125$
0.8776	0.8791	0.8817				0.8781
0.5403	0.5472	0.5605	0.5879			0.5423
0.0707	0.0845	0.1140				0.0746
−0.4151	−0.3977	−0.3552	−0.2608	−0.1907		0.4112
−0.8011	−0.7842	−0.7408				−0.7968
−0.9900	−0.9824	−0.9564	−0.8828			−0.9884
−0.9365	−0.9451	−0.9546				−0.9393
−0.6536	−0.6814	−0.7376	−0.8455	−1.059	−2.203	−0.6615
−0.2108	−0.2549	−0.3559				−0.2227
0.2837	0.2321	0.1031	−0.2152			0.2703
0.7087	0.6630	0.5354				0.6973
0.9602	0.9346	0.8436	0.5264	−0.1410		0.9546
0.9766	0.9820	0.9592				0.9793
0.7539	0.7942	0.8576	0.8494			0.7656
0.3466	0.4164	0.5634				0.3656
−0.1455	−0.0061	0.1444	0.5530	−0.7414	−0.1357	−0.1234
−0.6020	−0.5226	−0.3040				−0.5822
−0.9111	−0.8589	−0.6804	−0.1142			−0.8993
−0.9972	−0.9890	−0.9005				−0.9973
−0.8391	−0.8821	−0.9158	−0.6583	0.2635		−0.8525
−0.4755	−0.5641	−0.7244				−0.5002
0.0044	0.1114	0.3712	−0.7035			0.0026
0.4833	0.3675	0.0631				0.4543
0.8439	0.7578	0.4798	−0.2499	−0.4905	1.008	0.8239
0.9978	0.9661	0.7851				0.9929

quencies as degrees of freedom. For a conditionally stable numerical integration procedure to be usable, therefore, it would be necessary that the size of the time step be compatible with the highest natural frequency, even if the oscillations associated with it are of no interest. Otherwise, the amplitudes of those oscillations would grow without bound and wreak havoc on the whole procedure by polluting the lower-frequency oscillations as well. Unconditionally stable methods do not present this problem, since the higher frequency modes of vibration will get filtered out by the numerical damping if the size of the time step is too large to capture them. Perhaps the best way to illustrate these ideas is by means of a two-degree-of freedom system, as in the following example.

B.4 EXAMPLE: A TWO-DEGREES-OF-FREEDOM SYSTEM

We will examine and discuss the different behaviours of the central-difference and the Wilson θ-methods for the numerical analysis of the system with two degrees of freedom shown in Figure B.2. The masses are equal and the central spring is n times stiffer than the lateral ones.

In the absence of external forces, the differential equations of motion of the system are:

$$m\frac{d^2 x}{dx^2} + kx + nk(x - y) = 0 \qquad (B.49)$$

$$m\frac{d^2 y}{dt^2} + ky + nk(y - x) = 0 \qquad (B.50)$$

where x and y are the displacements of the masses, as shown in the figure. The general solution of this system of differential equations is:

$$x(t) = A \sin(\omega_1 t) + B \cos(\omega_1 t) + C \sin(\omega_2 t) + D \cos(\omega_2 t) \qquad (B.51)$$

$$y(t) = A \sin(\omega_1 t) + B \cos(\omega_1 t) - C \sin(\omega_2 t) - D \cos(\omega_2 t) \qquad (B.52)$$

where A, B, C and D are constants to be determined from the initial conditions and where ω_1 and ω_2 are the natural frequencies, given in this case by

$$\omega_1 = \left(\frac{k}{m}\right)^{1/2}, \qquad \omega_2 = \left[\frac{(2n+1)k}{m}\right]^{1/2} \qquad (B.53)$$

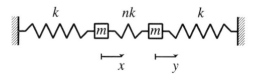

Figure B.2 Two-degrees-of-freedom system.

Topics in time-dependent modelling 213

We see, then, that the spread between the natural frequencies can be very wide according to how large the ratio between the spring stiffnesses is: for $n = 12$, for example, the second natural frequency is five times as large as the first. From the general solution we can also appreciate that to each natural frequency there corresponds a **natural mode of vibration**. In our case, the first natural frequency corresponds to both masses moving in unison by the same amount and in the same direction, while the motion associated with the second natural frequency consists of the masses moving in unison by the same amount but in opposite directions. Thus, the first natural mode, in this case, does not cause any stretch of the middle spring, which explains why its natural frequency coincides with that of the one-degree-of-freedom system of the previous example.

If we should start the motion of this system by applying equal initial values for x and y, with zero initial velocities, only the first natural mode would get excited and the system will keep oscillating *ad infinitum* in this first mode, without ever exciting the second. Indeed, let the initial values be $x(0) = y(0) = 1$, and $[dx/dt]_0 = [dy/dt]_0 = 0$. Introducing these values into the general solution (B.51)–(B.52), we immediately obtain $A = C = D = 0$ and $B = 1$, so that the solution corresponding to our initial conditions simply reads

$$x(t) = y(t) = \cos(\omega_1 t) \tag{B.54}$$

Let us now assume that a small perturbation, 10^{-5} say, has been given to the value of the initial displacement $y(0)$. This perturbation could be a true physical disturbance or a result of the round-off error of the computing device. Then, the second mode of vibration will be excited ever so slightly. If the numerical integration procedure being used is only conditionally stable, and if the step size has been chosen small enough to satisfy the stability condition for the lower frequency, but not quite small so as to satisfy it for the higher one, we will observe a growth of the second mode which, in a relatively short time, will obliterate any trace of accuracy in the solution.

Table B.3 shows a comparison of numerical results obtained both with the central-difference method and Wilson's θ-method. The results were obtained by means of two short *ad hoc* computer programs. Even at this level of simplicity (linearity, few degrees of freedom, etc.), the program for Wilson's θ-method is already much more involved than that for the central-difference method. Indeed, the former requires, at each step, the solution of a system of two simultaneous (linear) equations, while the latter consists of just a pair of explicit formulae. For this example we choose a perturbation of 10^{-5} on the initial value of y, a stiffness ratio $n = 12$, and two different values of the step, corresponding to $\alpha = 0.25$ and 0.50. The solution extends over two full periods.

We observe that the results of Wilson's θ-method are completely consistent with the fact that the second mode has practically no influence on the final result. Both masses undergo the same displacements at all times, and these are numerically identical to those of the corresponding one-degree-of-freedom system, as expected from the theory. The central-difference method, on the other hand,

Table B.3 Small perturbation of system with two degrees of freedom: comparative results

Central-difference method				Wilson's θ-method (with $\theta = 1.40$)			
$\alpha = 0.25$		$\alpha = 0.50$		$\alpha = 0.25$		$\alpha = 0.50$	
x	y	x	y	x	y	x	y
0.8770	0.8769	0.8750	0.8750	0.8791	0.8791	0.8817	0.8817
0.5381	0.5381	0.5312	0.5313	0.5472	0.5472	0.5605	0.5605
0.0668	0.0668	0.0549	0.0544	0.0846	0.0846	0.1139	0.1140
−0.4209	−0.4209	−0.4365	−0.4346	−0.3977	−0.3977	−0.3552	−0.3552
−0.8051	−0.8051	−0.8131	−0.8207	−0.7842	−0.7842	−0.7408	−0.7408
−0.9911	−0.9911	−1.0091	−0.9790	−0.9824	−0.9824	−0.9564	−0.9564
−0.9332	−0.9332	−0.8625	−0.9828	−0.9451	−0.9451	−0.9547	−0.9546
−0.6457	−0.6457	−0.8612	−0.3800	−0.6815	−0.6815	−0.7376	−0.7376
−0.1992	−0.1992	0.7991	−1.1259	−0.2550	−0.2549	−0.3559	−0.3559
0.2962	0.2962	−3.5154	4.1846	0.2321	0.2321	0.1031	0.1031
0.7188	0.7188	16.149	−14.651	0.6630	0.6630	0.5354	0.5354
0.9645	0.9645	−60.624	−62.576	0.9346	0.9346	0.8436	0.8436
0.9728	0.9728	247.36	−245.44	0.9820	0.9820	0.9592	0.9592
0.7417	0.7417	−984.90	986.30	0.7942	0.7942	0.8576	0.8576
0.3281	0.3281	3 943	−3942	0.4164	0.4164	0.5634	0.5634
−0.1663	−0.1662	−15 770	15 770	−0.0607	−0.0607	0.1444	0.1444
−0.6197	−0.6197	63 078	−63 079	−0.5226	−0.5226	−0.3040	−0.3040
−0.9206	−0.9206	−252 300	252 300	−0.8589	−0.8589	−0.6804	−0.6804
−0.9950	−0.9950	1 009 000	−1 009 000	−0.9890	−0.9890	−0.9005	−0.9005
−0.8245	−0.8245	−4 037 000	4 037 000	−0.8821	−0.8821	−0.9158	−0.9158
−0.4511	−0.4511	1.6×10^7	-1.6×10^7	−0.5641	−0.5641	−0.7244	−0.7244
0.0033	0.0033	-6.5×10^7	6.5×10^7	−0.1114	−0.1114	−0.3712	−0.3712
0.5095	0.5095	2.6×10^8	-2.6×10^8	0.3675	0.3675	0.0631	0.0631
0.8603	0.8603	-1.0×10^9	1.0×10^9	0.7578	0.7578	0.4798	0.4798
0.9994	0.9994	4.1×10^9	-4.1×10^9	0.9661	0.9661	0.7851	0.7851

Topics in time-dependent modelling 215

provides reliable results only when the step size is smaller than the limit imposed by the highest frequency, whether the corresponding mode of vibration is important or not. In our case, the limit corresponding to the higher frequency is $\alpha = 0.4$ (i.e. one-fifth of that for the lower frequency) and, correspondingly, the results for $\alpha = 0.5$ diverge rapidly, even though the initial conditions differ from the ideal ones only by one part in a hundred thousand! The results for $\alpha = 0.25$ are entirely consistent with the first mode and, moreover, coincide with their one-degree-of-freedom counterparts. Although this is a highly idealized example, muscle models will in general have too many degrees of freedom to risk the dangers of unwanted numerical instabilities. For this reason, the use of unconditionally stable implicit methods is recommended, in spite of their higher computational cost.

B.5 A PROGRAM FOR TIME-DEPENDENT ANALYSIS OF SKELETAL MUSCLE

The following is a complete listing of a program for the time-dependent analysis of skeletal muscle. The approximate force–velocity dependence of equation (6.6) has been incorporated in the constitutive routine, but it can be easily replaced by more sophisticated time-dependent models, such as those presented in Chapter 4.

```
DECLARE FUNCTION resid# (i%, x#(), vel#(), acc#(), n%)
DECLARE FUNCTION deriv# (i%, j%, fi0#, x#(), vel#(), acc#(), n%, h#, alph1#, bet1#)
DECLARE FUNCTION cons# (i%, j%, oldl#, newl#, rate#)
DECLARE SUB gauss (coef#(), rhs#(), deltax#(), n%)
DECLARE SUB houblt (max%, eps#, h#, n%, x0#(), vel0#(), acc0#(), alpha#(), beta#(), vel#())
DEFDBL A-Z
COMMON SHARED np%, mc%, xc(), forcec, ynf$, fibre(), bot(), top(), tendon, activ(), dt, mass(),
1r0
OPEN "c:dynamic.out" FOR OUTPUT AS #1
PRINT "This is a program for the dynamic solution of a straight-line planar"
PRINT " muscle model. Follow the self-explanatory input instructions."
PRINT " "
PRINT "        GENERAL APPEARANCE"
PRINT " "
PRINT " y-axis              o : node "
PRINT "  ^   2               f : fibre"
PRINT "  |   o   4            a : aponeurosis "
PRINT "  |    / a o"
PRINT "  |   /  / a 6"
PRINT "  |  f/   f/    o"
PRINT "  |  / panel /  f/  ."
PRINT "  | /  # 1 / panel /    .   2p             2p + 3"
PRINT "  | /     /   #2 /       o    2p + 2     (roller & force"
PRINT "  |/     /    /       f/   a (roller)     or fixed)"
```

Appendix B

```
PRINT "    o-------/-------/----------------/---------o-------------o------- >"
PRINT "    1   a   o       /                / panel /    tendon          x-axis"
PRINT " (fixed)  3       a o              / # p /f"
PRINT "              5    '    /         /"
PRINT "                       o    a    o"
PRINT "                            2p - 1    2p + 1"
'
'
INPUT "enter number of panels : ", np%
' calculate total number of nodes
mc% = 2 * np% + 3
DIM xc(mc%, 2), mass(mc%), fibre(np% + 1), bot(np%), top(np%), activ(np% + 1)
PRINT "next, you will enter the nodal coordinates in the x-y system depicted"
PRINT " (in mm) as well as the nodal masses (in grams)"
FOR i% = 1 TO mc%
PRINT "for node #", i%; : INPUT ; " x = ", xc(i%, 1): INPUT ; " y = ", xc(i%, 2): INPUT "
mass = ", mass(i%)
' convert masses to Megagrams
mass(i%) = mass(i%) * .000001
NEXT i%
PRINT "is the last node fixed, i.e., displacement controlled (y),"
INPUT "or is it movable, i.e., force controlled (n)? Enter y or n: ", ynf$
'calculate total number of unknowns
n% = 5 * (mc% - 1) / 2 + 1
PRINT "next you will enter the fibre-, aponeuroses- and tendon-areas of influence"
PRINT "                 (in square mm)"
FOR i% = 1 TO np%
PRINT "fibre "; 2 * i% - 1; "-"; 2 * i%; : INPUT ; ": ", fibre(i%): PRINT "   apo. "; 2 * i% - 1; "-";
2 * i% + 1; : INPUT ; ": ", bot(i%): PRINT " apo. "; 2 * i%; "-"; 2 * i% + 2; : INPUT ": ", top(i%)
NEXT i%
PRINT "fibre "; 2 * np% + 1; "-"; 2 * np% + 2; : INPUT ; ": ", fibre(np% + 1): INPUT " tendon:
", tendon
PRINT "next, you will enter the ratio : initial length to optimal length"
INPUT ; "(This ratio is assumed to be the same for all fibres) " , lr0
PRINT "next, you will enter parameters to control the numerical procedure:"
PRINT "   max = maximum number of iterations in Newton-Raphson (say, 50)"
PRINT "   eps = relative quadratic error for convergence (say, 0.005)"
PRINT "   h   = increment for numerical differentiation (say, 0.001) "
INPUT ; "max = ", max%: INPUT ; " eps = ", eps: INPUT " h = ", h
INPUT "Enter the time increment dt (in secs): ", dt
'set initial conditions to zero for first step
DIM x0(n%), vel0(n%), acc0(n%), vel(n%)
DIM xm1(n%), xm2(n%), alpha(4), beta(4)
FOR i% = 1 TO n%
x0(i%) = 0!
vel0(i%) = 0!
acc0(i%)=0!
NEXT i%
```

```
FOR i% = 1 TO n%
xm1(i%) = acc0(i%)
xm2(i%) = vel0(i%)
NEXT i%
' define the Houbolt constants for first step
alpha(1) = 3! / dt
alpha(2) = -3! / dt
alpha(3) = -.5 * dt
alpha(4) = -2!
beta(1) = 6! / dt ^ 2
beta(2) = -6! / dt ^ 2
beta(3) = -2!
beta(4) = -6! / dt
PRINT "Finally, you will enter the loading conditions"
tstep% = 0
77 '
tstep% = tstep% + 1
PRINT "STARTING INPUT FOR TIME STEP NO.", tstep%
IF ynf$ <> "n" THEN GOTO 81
INPUT "You have specified a movable end. Therefore, enter the force (in N):", forcec
GOTO 82
81 INPUT "You have specified controlled displacement. Enter it (in mm):", forcec
82 PRINT "You will now enter the fibre activation coefficients (between 0 and 1)"
FOR i% = 1 TO np% + 1
PRINT "fibre "; 2 * i% - 1; "-"; 2 * i%; : INPUT ": ", activ(i%)
NEXT i%
CALL houblt(max%, eps, h, n%, x0(), xm1(), xm2(), alpha(), beta(), vel())
PRINT #1, " "
PRINT #1, "SOLUTION FOR TIME STEP NO.", tstep%
PRINT #1, " applied";
IF ynf$ = "n" THEN PRINT #1, " force"; ELSE PRINT #1, " displacement";
PRINT #1, forcec
PRINT #1, " "
PRINT #1, "FIBRE SPECIFIED ACTIVATION COEFFICIENT"
FOR ip% = 1 TO np% + 1
PRINT #1, 2 * ip% - 1; 2 * ip%; activ(ip%)
NEXT ip%
PRINT #1, " "
PRINT #1, "NODE      X-DISPL      Y-DISPL"
FOR i% = 1 TO mc%
PRINT #1, i%, x0(2 * i% - 1), x0(2 * i%)
NEXT i%
' calculate internal forces
'
'
PRINT #1, " "
PRINT #1, "FIBRE    FORCE    APO    FORCE    APO    FORCE"
FOR ip% = 1 TO np% + 1
```

Appendix B

```
nn% = 2 * ip% - 1
jnn% = 2 * ip%
FOR ic% = 1 TO 3
oldl = 0#
newl = 0#
rate = 0#
FOR ks% = 1 TO 2
dij0 = xc(jnn%, ks%) - xc(nn%, ks%)
dij = dij0 + x0(2 * jnn% - 2 + ks%) - x0(2 * nn% - 2 + ks%)
rate = rate + dij * (vel(2 * jnn% - 2 + ks%) - vel(2 * nn% - 2 + ks%))
oldl = oldl + dij0 * dij0
newl = newl + dij * dij
NEXT ks%
oldl = SQR(oldl)
newl = SQR(newl)
rate = rate / newl
fff = cons(nn%, jnn%, oldl, newl, rate)
PRINT #1, nn%; jnn%; fff;
IF ip% = np% + 1 THEN GOTO 84
jnn% = jnn% + 1
IF jnn% = 2 * (ip% + 1) THEN nn% = nn% + 1
NEXT ic%
PRINT #1, " "
NEXT ip%
84 '
'
' tendon length
oldl = xc(mc%, 1) - xc(mc% - 1, 1)
newl = oldl + x0(2 * mc% - 1) - x0(2 * mc% - 3)
oldl = ABS(oldl)
newl = ABS(newl)
' rate is irrelevant for tendon
rate = 0!
PRINT #1, "calculated tendon force = ", cons(mc% - 1, mc%, oldl, newl, rate)
'update for next run
PRINT " continue? y or n": INPUT yn$: IF yn$ <> "y" THEN GOTO 88
PRINT "You have decided to continue. Present values will be used as initial"
PRINT "You will now enter the new loading. Follow instructions as before"
IF tstep% > 2 THEN GOTO 85
' define the Houbolt constants for second step
alpha(1) = 11! / (6 * dt)
alpha(2) = -16! / (6! * dt)
alpha(3) = 5! / (6! * dt)
alpha(4) = -dt / 3!
beta(1) = 2! / dt ^ 2
beta(2) = -4! / dt ^ 2
beta(3) = 2! / dt ^ 2
beta(4) = -1!
```

```
GOTO 87
' define the Houbolt constants for generic step
85 alpha(1) = 11! / (6 * dt)
alpha(2) = -3! / dt
alpha(3) = 1.5 / dt
alpha(4) = -1! / (3! * dt)
beta(1) = 2! / dt ^ 2
beta(2) = -5! / dt ^ 2
beta(3) = 4! / dt ^ 2
beta(4) = -1! / dt ^ 2
87 '
GOTO 77
88 '
END



FUNCTION cons (i%, j%, oldl, newl, rate)
' Available through COMMON SHARED:
'np%, mc%, xc(), forcec, ynf$, fibre(), bot(), top(), tendon, activ(), dt, mass(), lr0
aux = 0!
' distinguish between fibres and aponeuroses (or tendon)
big% = i%
small% = j%
IF small% < big% GOTO 109
big% = j%
small% = i%
109 dif% = big% - small%
ON (dif% + 1) GOTO 117, 111, 115, 117
111 IF big% = mc% THEN GOTO 114
IF small% = 2 * INT(small% / 2) THEN GOTO 117
' fibre constitutive equation
veloc = -rate
optl = oldl / lr0
lratio = newl / optl
'force-length influence
fl = -2.78 * lr0 ^ 2 + 5.56 * lr0 - 1.78
'force-velocity influence
ppp = EXP(veloc / 120!)
fv = 1 - ((ppp - 1! / ppp) / (ppp + 1! / ppp))
aux =  fv * activ(big% / 2) * fibre(big% / 2) * (5! / 18!) * (fl + 0.9664 * (lratio - lr0))
GOTO 117
114 area = tendon
GOTO 116
115 panel% = INT((big% - 1) / 2)
area = bot(panel%)
IF big% = 2 * (panel% + 1) THEN area = top(panel%)
' tendon-like constitutive equation
116 strain = (newl - oldl) / oldl
```

Appendix B

```
' force in newtons when area in square mm
aux = (15160! * ((strain + .000725) ^ 2 - 5.26E-07)) * area
' extend linearly to the (unphysical) compressive range
IF strain < 0 THEN aux = 15160! * 2! * .000725 *strain * area
117 cons = aux
END FUNCTION


FUNCTION deriv (i%, j%, fi0, x(), vel(), acc(), n%, h, alph1, bet1)
' ********************************************************************************
' This function calculates a derivative numerically by forward differences.
' To save computing effort, it is assumed that the value of the function at
' the original point has already been calculated: it is passed as argument
' fi0. h is a given non-zero increment. Ideally, this h could be varied
' for different kinds of variables.
' ********************************************************************************
' version re-adapted to Houbolt's method
' available through COMMON SHARED: dt
'th = 1.4
xx = x(j%)
vvel = vel(j%)
aacc = acc(j%)
x(j%) = x(j%) + h
deltav = alph1 * h
vel(j%) = vel(j%) + deltav
deltac = bet1 * h
acc(j%) = acc(j%) + deltac
fi0h = resid(i%, x(), vel(), acc(), n%)
x(j%) = xx
vel(j%) = vvel
acc(j%) = aacc
deriv = (fi0h - fi0) / h
END FUNCTION


SUB gauss (coef(), rhs(), soln(), n%)
DIM aux(n%, n%)
FOR i% = 1 TO n%
soln(i%) = rhs(i%)
FOR j% = 1 TO n%
aux(i%, j%) = coef(i%, j%)
NEXT j%
NEXT i%
IF n% = 1 THEN GOTO 58
FOR k% = 2 TO n%
IF aux(k% - 1, k% - 1) <> 0! THEN GOTO 57
FOR l% = k% TO n%
IF aux(l%, k% - 1) = 0! THEN GOTO 55
FOR m% = k% - 1 TO n%
x = aux(k% - 1, m%)
```

```
aux(k% - 1, m%) = aux(l%, m%)
aux(l%, m%) = x
NEXT m%
x = soln(k% - 1)
soln(k% - 1) = soln(l%)
soln(l%) = x
GOTO 57
55 ' keep looking for non-zero pivot
NEXT l%
56 PRINT "singularity detected by Gauss"
STOP
57 ' successful non-zero pivot search. Equations exchanged if necessary.
FOR i% = k% TO n%
x = aux(i%, k% - 1) / aux(k% - 1, k% - 1)
soln(i%) = soln(i%) - x * soln(k% - 1)
FOR j% = k% TO n%
aux(i%, j%) = aux(i%, j%) - x * aux(k% - 1, j%)
NEXT j%
NEXT i%
NEXT k%
58 IF aux(n%, n%) = 0! THEN GOTO 56
soln(n%) = soln(n%) / aux(n%, n%)
IF n% = 1 THEN GOTO 59
FOR k% = n% - 1 TO 1 STEP -1
FOR j% = k% + 1 TO n%
soln(k%) = soln(k%) - aux(k%, j%) * soln(j%)
NEXT j%
soln(k%) = soln(k%) / aux(k%, k%)
NEXT k%
59 ' Solution completed
END SUB

SUB houblt (max%, eps, h, n%, x0(), xm1(), xm2(), alpha(), beta(), vel())
' ******************************************************************************
' Newton-Raphson nonlinear solver adapted to Houbolt's method
' max% = maximum number of iterations
' eps = tolerance
' h = increment for (possibly) numerical differentiation (see SUB deriv)
' n% = number of equations and unknowns
' x0() = initial solution at time t
' xm1() = solution at time t-h (or, for first step, initial acceleration)
' xm2() = soln at time t-2h (1st step: init. veloc., 2nd: init. accel.)
' deltax() = Newton-Raphson-generated increments of x0()
' x(), vel(), acc() = updated solution at t + h
' available through COMMON SHARED: mc% = number of nodes, dt = time incr.
' Will return x in x0, x0 in xm1, xm1 in xm2
' ******************************************************************************
DIM coef(n%, n%), rhs(n%), deltax(n%), x(n%), acc(n%)
' copy initial to present values for a start
```

```
FOR i% = 1 TO n%
x(i%) = x0(i%)
NEXT i%
FOR k% = 1 TO max%
    FOR i% = 1 TO n%
        vel(i%) = alpha(1) * x(i%) + alpha(2) * x0(i%) + alpha(3) * xm1(i%) + alpha(4) * xm2(i%)
        acc(i%) = beta(1) * x(i%) + beta(2) * x0(i%) + beta(3) * xm1(i%) + beta(4) * xm2(i%)
    NEXT i%
    ' Prepare right-hand side vector rhs() and coefficient matrix coef()
    FOR i% = 1 TO n%
        rhs(i%) = -resid(i%, x(), vel(), acc(), n%)
        FOR j% = 1 TO n%
            coef(i%, j%) = deriv(i%, j%, -rhs(i%), x(), vel(), acc(), n%, h, alpha(1), beta(1))
        NEXT j%
    NEXT i%
    ' Solve system of linear equations
    CALL gauss(coef(), rhs(), deltax(), n%)
    ' Update solution and calculate quadratic error
    sum1 = 0!
    sum2 = 0!
    FOR i% = 1 TO n%
        x(i%) = x(i%) + deltax(i%)
        sum1 = sum1 + x(i%) * x(i%)
        sum2 = sum2 + deltax(i%) * deltax(i%)
    NEXT i%
    IF (SQR(sum2 / sum1) <= eps) THEN GOTO 99
NEXT k%
PRINT "Convergence not achieved in Newrap after "; max%; " iterations"
STOP
99 ' Convergence achieved. Update initial values for next time step
FOR i% = 1 TO n%
    vel(i%) = alpha(1) * x(i%) + alpha(2) * x0(i%) + alpha(3) * xm1(i%) + alpha(4) * xm2(i%)
    xm2(i%) = xm1(i%)
    xm1(i%) = x0(i%)
    x0(i%) = x(i%)
NEXT i%
END SUB


FUNCTION resid (i%, x(), vel(), acc(), n%)
' x, vel and acc are present values of displ, veloc and acceleration.
' Available through COMMON SHARED:
'                   np% = number of panels
'                   mc% = number of nodes
'                   xc(j%,k%) = initial kth coord of jth node
'                   forcec = applied force (if any)
'                   ynf$ = "y" if right end is fixed, else = "n"
'                   fibre() = fibre areas
'                   bot() = lower aponeurosis areas
```

Topics in time-dependent modelling 223

```
'                   top() = upper aponeuroses areas
'                   tendon = tendon area
'                   activ() = fibre activation coefficients
'                   mass() = nodal masses
' Support conditions enforced directly through equations u = 0.
' Only panel constraints are treated by means of Lagrange multipliers.
'NOTE: For this version of RESID, MAIN should specify n% = 5*(mc% - 1)/2 + 1
IF i% > 2 GOTO 11
sum = x(i%)
GOTO 49
11 IF i% >= 2 * (mc% - 1) GOTO 44
sum = 0#
'find left-panel number
pn% = INT((i% - 1) / 4)
pnl% = pn%
pnr% = pn% + 1
IF pnl% = 0 THEN pnl% = pnr%
IF pnl% = np% THEN pnr% = pnl%
'influence of panels on node (constraint equations times multiplier)
FOR jp% = pnl% TO pnr%
v1x = xc(2 * jp% + 2, 1) - xc(2 * jp% - 1, 1)
v1y = xc(2 * jp% + 2, 2) - xc(2 * jp% - 1, 2)
v2x = xc(2 * jp%, 1) - xc(2 * jp% + 1, 1)
v2y = xc(2 * jp%, 2) - xc(2 * jp% + 1, 2)
v1x = v1x + x(4 * jp% + 3) - x(4 * jp% - 3)
v1y = v1y + x(4 * jp% + 4) - x(4 * jp% - 2)
v2x = v2x + x(4 * jp% - 1) - x(4 * jp% + 1)
v2y = v2y + x(4 * jp%) - x(4 * jp% + 2)
ON (i% - 4 * jp% + 4) GOTO 31, 32, 33, 34, 35, 36, 37, 38
31 sum = sum + .5 * (-v2y) * x(2 * mc% + jp%)
GOTO 39
32 sum = sum + .5 * (v2x) * x(2 * mc% + jp%)
GOTO 39
33 sum = sum + .5 * (-v1y) * x(2 * mc% + jp%)
GOTO 39
34 sum = sum + .5 * (v1x) * x(2 * mc% + jp%)
GOTO 39
35 sum = sum + .5 * (v1y) * x(2 * mc% + jp%)
GOTO 39
36 sum = sum + .5 * (-v1x) * x(2 * mc% + jp%)
GOTO 39
37 sum = sum + .5 * (v2y) * x(2 * mc% + jp%)
GOTO 39
38 sum = sum + .5 * (-v2x) * x(2 * mc% + jp%)
39 '
NEXT jp%
'equilibrium influences
'find node number
```

```
nn% = INT((i% + 1) / 2)
'find parity
k% = i% - 2 * INT((i% - 1) / 2)
FOR jn% = nn% - 2 TO nn% + 2 STEP 2
IF jn% <=  0 OR jn% >=  mc% THEN GOTO 43
jnn% = jn%
IF jn% <> nn% GOTO 41
jnn% = jn% + 1
IF nn% = 2 * INT(nn% / 2) THEN jnn% = jn% - 1
41 aux = 0#
oldl = 0#
newl = 0#
rate = 0#
FOR ks% = 1 TO 2
dij0 = xc(jnn%, ks%) - xc(nn%, ks%)
dij = dij0 + x(2 * jnn% - 2 + ks%) - x(2 * nn% - 2 + ks%)
oldl = oldl + dij0 * dij0
newl = newl + dij * dij
rate = rate + dij * (vel(2 * jnn% - 2 + ks%) - vel(2 * nn% - 2 + ks%))
IF ks% = k% THEN aux = aux + dij
NEXT ks%
oldl = SQR(oldl)
newl = SQR(newl)
rate = rate / newl
sum = sum - cons(nn%, jnn%, oldl, newl, rate) * aux / newl
43 NEXT jn%
sum = sum + mass(nn%) * acc(i%)
IF i% <> 2 * mc% - 3 THEN GOTO 49
' tendon treatment
oldl = xc(mc%, 1) - xc(mc% - 1, 1)
newl = oldl + x(2 * mc% - 1) - x(2 * mc% - 3)
oldl = ABS(oldl)
newl = ABS(newl)
'note: rate is irrelevant for tendon
rate = vel(2 * mc% - 1) - vel(2 * mc% - 3)
sum = sum - cons(mc% - 1, mc%, oldl, newl, rate)
GOTO 49
44 IF i% > 2 * mc% GOTO 46
sum = x(i%)
IF i% = 2 * mc% - 1 THEN sum = sum - forcec
IF ynf$ <> "n" OR i% <> 2 * mc% - 1 THEN GOTO 49
oldl = xc(mc%, 1) - xc(mc% - 1, 1)
newl = oldl + x(2 * mc% - 1) - x(2 * mc% - 3)
oldl = ABS(oldl)
newl = ABS(newl)
rate = vel(2 * mc% - 1) - vel(2 * mc% - 3)
sum = cons(mc% - 1, mc%, oldl, newl, rate) - forcec + mass(mc%) * acc(2 * mc% - 1)
GOTO 49
```

Topics in time-dependent modelling 225

```
46 ' panel constraint equations
   ' calculate areas as cross-product of panel diagonals
   'find panel number
   pn% = i% - 2 * mc%
   vlx = xc(2 * pn% + 2, 1) - xc(2 * pn% - 1, 1)
   vly = xc(2 * pn% + 2, 2) - xc(2 * pn% - 1, 2)
   v2x = xc(2 * pn%, 1) - xc(2 * pn% + 1, 1)
   v2y = xc(2 * pn%, 2) - xc(2 * pn% + 1, 2)
   area = .5 * (vlx * v2y - vly * v2x)
   vlx = vlx + x(4 * pn% + 3) - x(4 * pn% - 3)
   vly = vly + x(4 * pn% + 4) - x(4 * pn% - 2)
   v2x = v2x + x(4 * pn% - 1) - x(4 * pn% + 1)
   v2y = v2y + x(4 * pn%) - x(4 * pn% + 2)
   area1 = .5 * (vlx * v2y - vly * v2x)
   sum = area1 - area
49 resid = sum
END FUNCTION
```

B.6 EXAMPLE: TIME-DEPENDENT DEFORMATION OF A CAT MEDIAL GASTROCNEMIUS MUSCLE

This example has already been presented in Chapter 6. Here, we reproduce verbatim the corresponding interactive computer session.

This is a program for the dynamic solution of a straight-line planar muscle model. Follow the self-explanatory input instructions.

```
            GENERAL APPEARANCE

y-axis                       o : node
  ^    2                     f : fibre
  |    o     4               a : aponeurosis
  |   / a   o
  |  /    / a   6
  | f/   f/     o
  | / panel /   f/  .
  | / # 1  / panel /     . 2p                    2p + 3
  | /     / # 2  /          o      2p + 2       (roller & force
  |/     /      /          f/     a (roller)     or fixed)
  o--------/---------/-------------/----------o----------------------------o------ >
  1  a   o         /       / panel /      tendon              x-axis
 (fixed) 3   a   o         /    # p /f
             5       '    /       /
                         o   a    o
                        2p - 1   2p + 1

enter number of panels : 3
```

Appendix B

next, you will enter the nodal coordinates in the x-y system depicted
(in mm) as well as the nodal masses (in grams)
for node # 1 x = 0 y = 0 mass = 0
for node # 2 x = 18.1 y = 8.5 mass = 0
for node # 3 x = 19.7 y = -3.5 mass = 0
for node # 4 x = 41 y = 5.1 mass = 0
for node # 5 x = 39.4 y = -6.9 mass = 0
for node # 6 x = 64.1 y = 1.1 mass = 0
for node # 7 x = 59 y = -10.4 mass = 0
for node # 8 x = 88 y = 0 mass = 0
for node # 9 x = 120 y = 0 mass = 0
is the last node fixed, i.e., displacement controlled (y),
or is it movable, i.e., force controlled (n)? Enter y or n: n
next you will enter the fibre-, aponeuroses- and tendon-areas of influence
          (in square mm)
fibre 1 - 2 : 75   apo. 1 - 3 : 2 apo. 2 - 4 : 4
fibre 3 - 4 : 150 apo. 3 - 5 : 3 apo. 4 - 6 : 3
fibre 5 - 6 : 150 apo. 5 - 7 : 4 apo. 6 - 8 : 2
fibre 7 - 8 : 75   tendon: 4
next, you will enter the ratio: initial length to optimal length
(This ratio is assumed to be the same for all fibres) lr0 = 0.8
next, you will enter parameters to control the numerical procedure:
 max = maximum number of iterations in Newton-Raphson (say, 50)
 eps = relative quadratic error for convergence (say, 0.005)
 h   = increment for numerical differentiation (say, 0.001)
max = 100 eps = 0.001 h = 0.001
Enter the time increment dt (in secs): 0.05
Finally, you will enter the loading conditions
STARTING INPUT FOR TIME STEP NO.   1
You have specified a movable end. Therefore, enter the force (in N):10
You will now enter the fibre activation coefficients (between 0 and 1)
fibre 1 - 2 : .1
fibre 3 - 4 : .1
fibre 5 - 6 : .1
fibre 7 - 8 : .1
 continue? y or n
? y
You have decided to continue. Present values will be used as initial
You will now enter the new loading. Follow instructions as before
STARTING INPUT FOR TIME STEP NO.   2
You have specified a movable end. Therefore, enter the force (in N):20
You will now enter the fibre activation coefficients (between 0 and 1)
fibre 1 - 2 : .2
fibre 3 - 4 : .2
fibre 5 - 6 : .2
fibre 7 - 8 : .2
 continue? y or n
? y

Topics in time-dependent modelling 227

```
You have decided to continue. Present values will be used as initial
You will now enter the new loading. Follow instructions as before
STARTING INPUT FOR TIME STEP NO.          3
You have specified a movable end. Therefore, enter the force (in N):50
You will now enter the fibre activation coefficients (between 0 and 1)
fibre 1 - 2 : .6
fibre 3 - 4 : .6
fibre 5 - 6 : .6
fibre 7 - 8 : .6
  continue? y or n
? y
You have decided to continue. Present values will be used as initial
You will now enter the new loading. Follow instructions as before
STARTING INPUT FOR TIME STEP NO.          4
You have specified a movable end. Therefore, enter the force (in N):100
You will now enter the fibre activation coefficients (between 0 and 1)
fibre 1 - 2 : 1
fibre 3 - 4 : 1
fibre 5 - 6 : 1
fibre 7 - 8 : 1
  continue? y or n
? y
You have decided to continue. Present values will be used as initial
You will now enter the new loading. Follow instructions as before
STARTING INPUT FOR TIME STEP NO.          5
You have specified a movable end. Therefore, enter the force (in N):100
You will now enter the fibre activation coefficients (between 0 and 1)
fibre 1 - 2 : 1
fibre 3 - 4 : 1
fibre 5 - 6 : 1
fibre 7 - 8 : 1
  continue? y or n
? y
You have decided to continue. Present values will be used as initial
You will now enter the new loading. Follow instructions as before
STARTING INPUT FOR TIME STEP NO.          6
You have specified a movable end. Therefore, enter the force (in N):100
You will now enter the fibre activation coefficients (between 0 and 1)
fibre 1 - 2 : 1
fibre 3 - 4 : 1
fibre 5 - 6 : 1
fibre 7 - 8 : 1
  continue? y or n
? y
You have decided to continue. Present values will be used as initial
You will now enter the new loading. Follow instructions as before
STARTING INPUT FOR TIME STEP NO.          7
You have specified a movable end. Therefore, enter the force (in N):100
```

You will now enter the fibre activation coefficients (between 0 and 1)
fibre 1 - 2 : 1
fibre 3 - 4 : 1
fibre 5 - 6 : 1
fibre 7 - 8 : 1
 continue? y or n
? y
You have decided to continue. Present values will be used as initial
You will now enter the new loading. Follow instructions as before
STARTING INPUT FOR TIME STEP NO.   8
You have specified a movable end. Therefore, enter the force (in N):100
You will now enter the fibre activation coefficients (between 0 and 1)
fibre 1 - 2 : 1
fibre 3 - 4 : 1
fibre 5 - 6 : 1
fibre 7 - 8 : 1
 continue? y or n
? n

SOLUTION FOR TIME STEP NO. 1   (time = .05)
 applied force 10

FIBRE SPECIFIED ACTIVATION COEFFICIENT
 1 2 .1
 3 4 .1
 5 6 .1
 7 8 .1

 NODE   X-DISPL   Y-DISPL
  1   0   0
  2   .2862268759980755   .7689743655006281
  3   .4639275969653942   .9448905996609269
  4   .4179712737770098   .7673758326492045
  5   .6529898037787032   .8499353646416808
  6   .65644370419689   .7322439504424891
  7   .7620883772603124   .9146220310335228
  8   .996679253871194   0
  9   1.385066621193349   0

FIBRE   FORCE   APO   FORCE   APO   FORCE
 1 2 2.454377359323193 1 3 8.290677985745038 2 4 2.42434697801494
 3 4 3.488916470213995 3 5 5.336974330934583 4 6 5.482828535535811
 5 6 3.639191770126769 5 7 1.840497714949774 6 8 8.519615742570299
 7 8 1.793206322577473 calculated tendon force =   10.00000600074977

Topics in time-dependent modelling

```
SOLUTION FOR TIME STEP NO. 2    (time = .1 )
applied force 20


FIBRE SPECIFIED ACTIVATION COEFFICIENT
1 2 .2
3 4 .2
5 6 .2
7 8 .2



NODE      X-DISPL           Y-DISPL
 1     0          0
 2     .5119130423989936    1.181834030590843
 3     .6665294071562856    1.43207088556319
 4     .6569088692021602     .998875884247192
 5     .9149319300785694    1.126362268907814
 6     .9789495319517821     .8625549785690942
 7    1.062940667301957     1.119511695808802
 8    1.471606341005348     0
 9    2.030015873142842     0



FIBRE    FORCE        APO    FORCE         APO    FORCE
1 2  3.996181845811614  1 3  17.23434779948027  2 4  3.950981992325965
3 4  7.338864114551856  3 5  11.11947484440481  4 6  10.56423590699166
5 6  7.4668507423484    5 7   3.950681865907861 6 8  16.80464335870422
7 8  3.851077288183446  calculated tendon force =    20.00000904918226



SOLUTION FOR TIME STEP NO. 3    (time = .15 )
applied force 50


FIBRE SPECIFIED ACTIVATION COEFFICIENT
1 2 .6
3 4 .6
5 6 .6
7 8 .6



NODE      X-DISPL           Y-DISPL
 1     0          0
 2     .1509924007064769   1.201152248040949
 3     .9515351304534703   1.477139679213759
 4     .4044846680238542   1.037487645493293
 5    1.380889463133074    1.142814190019251
 6     .9373304987305646    .909470183886864
```

```
7     1.637130389277291    1.198015272364775
8     1.752479141980623    0
9     2.648445042158714    0


FIBRE    FORCE         APO    FORCE          APO    FORCE
1  2  9.741480774722445  1  3  43.31301339343108  2  4  9.608052557191925
3  4  18.53197355956013  3  5  27.97926267127536  4  6  26.28883513805533
5  6  18.66970242814714  5  7  10.08964176220069  6  8  41.82228455316451
7  8  9.847299148114836  calculated tendon force =   50.00000892536851



SOLUTION FOR TIME STEP NO. 4     (time = .2 )
applied force 100


FIBRE SPECIFIED ACTIVATION COEFFICIENT
1  2  1
3  4  1
5  6  1
7  8  1


NODE     X-DISPL              Y-DISPL
1     0              0
2     8.365594379023251D-02    1.086972640156853
3     1.259909291250892        1.436042932211677
4     .474893322508383         1.001364369507402
5     1.905693850682905        1.148838290435183
6     1.249416856680571        .901044134690301
7     2.279717767612957        1.264592669378255
8     2.434044948394699        0
9     3.710534532270868        0


FIBRE    FORCE         APO    FORCE          APO    FORCE
1  2  19.87045656180762  1  3  86.25554627047454  2  4  19.58243076172056
3  4  36.75720795687211  3  5  55.80821484259906  4  6  52.57931247782245
5  6  37.23452910364333  5  7  20.13438255756519  6  8  83.59108657696976
7  8  19.66909263129203  calculated tendon force =   100.0000021377431


SOLUTION FOR TIME STEP NO. 5     (time = .25 )
applied force 100


FIBRE SPECIFIED ACTIVATION COEFFICIENT
1  2  1
3  4  1
```

5 6 1
7 8 1

NODE   X-DISPL   Y-DISPL
1   0   0
2   .1917086284777829   1.095915510879375
3   1.262187975785886   1.454579246013304
4   .5739081495031007   .9603935257286171
5   1.898183075756268   1.123323945965254
6   1.343928250244015   .8203203609762364
7   2.257252271188475   1.193644746272602
8   2.539761566000323   0
9   3.81625156939184   0
FIBRE   FORCE   APO   FORCE   APO   FORCE
1 2 19.71723512188528 1 3 86.32984305187296 2 4 19.44171202775105
3 4 37.1584525092999 3 5 55.53304948005048 4 6 52.88632668538501
5 6 37.65856774749196 5 7 19.35344234881546 6 8 84.2421347316719
7 8 18.89441030481596 calculated tendon force = 100.0000667145834

SOLUTION FOR TIME STEP NO. 6   (time = .3)
applied force 100

FIBRE SPECIFIED ACTIVATION COEFFICIENT
1 2 1
3 4 1
5 6 1
7 8 1

NODE   X-DISPL   Y-DISPL
1   0   0
2   .3827444096781334   1.082796558711181
3   1.263056655344452   1.466191315252874
4   .7644122137734987   .9473478921263484
5   1.900559026852933   1.144653676279152
6   1.537399613376694   .8278078263924822
7   2.262189333544605   1.22364474616974
8   2.73164590559252   0
9   4.008136493963203   0

FIBRE   FORCE   APO   FORCE   APO   FORCE
1 2 19.64501466843797 1 3 86.28781024410914 2 4 19.39136064001799
3 4 37.04379175659639 3 5 55.49030117600114 4 6 52.81007532076545
5 6 37.476966711281 5 7 19.46881651245323 6 8 84.1072431315762
7 8 19.00769290856433 calculated tendon force = 100.0001567616374

SOLUTION FOR TIME STEP NO. 7   (time = .35)
applied force 100

FIBRE SPECIFIED ACTIVATION COEFFICIENT
1 2 1
3 4 1
5 6 1
7 8 1

NODE	X-DISPL	Y-DISPL
1	0	0
2	.551559283467802	1.066596984298147
3	1.263143404597954	1.47158466199826
4	.9333155757841622	.9316601505079105
5	1.901878900782593	1.1593166142763
6	1.708768842366667	.8278139269953991
7	2.264715337419988	1.243364962142607
8	2.902537796279021	0
9	4.179028151181832	0

FIBRE   FORCE   APO   FORCE   APO   FORCE
1 2 19.62837178449317 1 3 86.21768469029172 2 4 19.39235890762842
3 4 36.94176058394196 3 5 55.41824429407345 4 6 52.77919633570749
5 6 37.3435784346579 5 7 19.5054498551384 6 8 84.04377802803556
7 8 19.04322616386934 calculated tendon force =   100.0001208234398

SOLUTION FOR TIME STEP NO. 8   (time = .4)
applied force 100

FIBRE SPECIFIED ACTIVATION COEFFICIENT
1 2 1
3 4 1
5 6 1
7 8 1

NODE	X-DISPL	Y-DISPL
1	0	0
2	.6721095191308969	1.0552340365091
3	1.263177415980389	1.475146289829481
4	1.053180848018676	.9155488600394645
5	1.901649107863356	1.16456252329157
6	1.829868360380401	.8177219144724991

```
7    2.263502586931088    1.24693350720734
8    3.024724716657203    0
9    4.301214620821282    0


FIBRE   FORCE            APO   FORCE               APO   FORCE
1  2  19.61981779957154   1  3  86.16776925747138   2  4  19.39574112838137
3  4  36.91800114195605   3  5  55.32901931539994   4  6  52.80531133297036
5  6  37.30397133687237   5  7  19.43052896982713   6  8  84.08558331396424
7  8  18.96856078014753  calculated tendon force =   100.000051440303
```

REFERENCES

[1] Houbolt, J.C. (1950) A recurrence matrix solution for the dynamic response of elastic aircraft. *J. Aeronautical Sci.*, **17**: 540–550.
[2] Wilson, E.L., Farhoomand, I. and Bathe, K.J. (1973) Nonlinear dynamic analysis of complex structures. *Int. J. Earthquake Engng Str. Dynamics*, **1**: 241–252.

Index

Note: Page numbers in *italics* refer to Figures

A- (anisotropic) band   5
acetylcholine (Ach)   13
acetylcholinesterase   13
actin   4
action potential   12, 13
afferent axons   9
anelastic materials   87
angle of pinnation   23, *23*, 48–52, *49*, *50*
ankle joint   8
artificially stimulated contractions   9–10
ascending limb of force–length relationship   28
ATP hydrolysis reaction   16–17, 18

bipennate muscles   8, *8*, 25

carbohydrates   17
cat medial gastrocnemius muscle
  size principle   11
  static analysis program for   148–50, 195–199, 199–201
  time-dependent analysis program for   152, 225–33
Cauchy–Green tensor of continuum mechanics   123
central differences   202
  dynamics of one-degree-of-freedom system by   204–6
  two-degrees-of-freedom system   212–15, *212*
characteristic curve   104, *105*
compression   126
concentric contractions   18
conditional stability   203, 205
constitutive equation   127–8
constitutive theory   115, 118
constraints *see* geometric constraints

contractile element (CE)   89
  properties   25–47
    force–length properties   26–36, *29*
    force–velocity properties   26, 36–43, *36*, 40, *90*
    history-dependent properties   43–7
contraction, definition   3
convergence criterion   151
cross-bridge, definition   5, 6
cross-bridge cycle   14–16, *14*, *15*
cross-bridge model (Huxley)   70, 75–84, 101–5, *102*
  1957 formulation   77–80, *78*
  1971 formulation   80–4
  attachment/detachment probability distributions   105–6, *106*
  initial manifold   107–8, *107*
  macroscopic quantities and   106–9
cross-bridge theory   19–20
current position   120

d'Alembert's principle   129
degrees of freedom   114, *116*
descending limb force–length relationship   29
discretization   151
displacement   118–23
displacement vector   120, *120*
dynamics   115, 117–18

eccentric contractions   18
efferent axons   9
elastic fibres   127
elbow, flexion–extension at   158–9, *158*
elongation   118–23
endergonic reactions   16
endomysium   3

Index

energy balance during muscular contraction 18
epimysium 3
equilibrium equations
  computer code for 184–7
  for planar bar assembly 134–6
  for preservation of area 138–44
excitation–contraction coupling 11–16
exergonic reactions 16
external forces 124–5
external virtual work 129

fascia 3
fascicle 3
fatty acid oxidation 17
finite-difference method 202
force–length properties 26–36, *29*
  determination 34
force production 19–20
force sharing 167–70
force–velocity properties 26, 36–43, *36*, 40, *90*
forces of inertia 129
four-element model *110*
free-body equilibrium diagram 142, *142*, *143*
fused tetanic contractions 9
fusiform muscles 8, *8*

gastrocnemius muscles 8, 160
  angle of pinnation 49–50
  cat
    size principle 11
    static analysis program for 148–50, 195–199, 199–201
    time-dependent analysis program for 152, 225–33
Gauss elimination procedure 174, 179, 180
geometric constraints 115, 136–8
  of assemblage 115–16
  computer code for 184–7
  external supports 116, *117*
  preservation of panel area 116, *117*
geometrical nonlinearity 117
geometrically linearized theory 122
glucose breakdown 17
glycogen breakdown 17
gracilis muscle 160

Hill, Archibald Vivian 70
Hill type models 70–5
  limitations 74–5

Hill's equation of force–velocity relation 36–7, 38, 39, 40, 89
Hill's three-element model 89–91
  critique and possible extensions 91–3
  examples
    effects of speed and lingering time 95–6
    protocols of stretch 93–4
  history-dependent properties 43–7
    force depressions following shortening 43–5, *44*
    force enhancement following stretch 45–6, *46*
    stretch-shortening versus shortening-stretch 46–7, *47*
Houbolt's method 152, 207–9
Huxley, Andrew Fielding 70, 75
Huxley type models *see* cross-bridge model
Huxley–Zahalak equation 104, 108
Huxley's equation 102

I- (isotropic) band 5
inertia, forces of 129
initial position 120
instability, definition 29–30
internal forces 124–5
internal virtual work 129
inverse square law 128
isokinetic contractions 19
isometric contractions 18
isotonic contractions 19

joints 116

Kelvin body 88, *110*
kinematics 115–16
  of a fibre 120, *121*
knee joint 8
  flexion–extension at 159–60, *159*

lactic acid 17
Lagrange multipliers, method of 137, 142, 143, 145, 182, 184, 185
law of universal gravitational attraction 128
left-handed frame 119
linear dashpot 87, *88*
linear elastic element 87
linear equations, computer code for solving 179–81

linear spring   87, *87*
linear viscoelastic element   87

main processor   146
material system   124
maximal contraction   9
Maxwell body   88, *109*
memory effects   93
mesh points   202
minimum-fatigue criterion   165
modified axial force   130
α motor neurones   9
  action potential   12
α-γ motor neurone system   155–7
motor unit   9–11
movement control
  anatomy   157–60
  future research   171–2
  neurophysiology   153–7
  theoretical and experimental
    considerations   160–70
multipennate muscles   8, *8*
muscle fatigue   17
muscle fibre   3
muscle shapes   8
muscle spindle lengths   24
myofibrils   3
myofilaments, compliant   20
myosin   4

natural frequency   204
natural mode of vibration   213
natural period   205
neural pathways   153–7
neuromuscular junction   12
Newton's Second Law   117
Newton's Third Law   124–5
Newton–Raphson method   174
  computer code for   176–7
nodes   116
non-commutativity   93
nonlinear problems   174–6
numerical stability   205

observer   118
observer indifference   127
one-degree-of-freedom system   131–4
  dynamics of   204–6

palmitate   17
panel area preservation   145
parallel arrangement   88, *88*

parallel-fibre muscle   *106*
parallel-fibred models   24–5
passive element properties   52–7
  in parallel with contractile
    element   52–3
  in series with contractile
    element   53–7
  aponeuroses   55–7
  tendons   53–5, *56*
pawshake response   161–2
pennate muscles   8
pennate-fibred models   24–5
perimysium   3
phosphocreatine   16–17
physiological cross-sectional area
    (PCSA)   26
Piola–Kirchhoff stress tensor   130
plantaris muscles   8, 160
plateau region force–length
    relationship   28
popliteus   160
position vector   118
post-processor   146
postsynaptic terminal   13
power–velocity properties   26, 39–40, *41*
pre-processor   146
presynaptic terminal   12
principle of action and reaction (Newton's
    Third Law)   124–5
principle of virtual law   128–31

quadriceps muscle group   35, *35*

rectus femoris   159
  force–length properties   32–3
  working range   31, *34*
redundant equations, solving   161–4
reference frame   118, *119*
reference position   120
residuals   175
rheological models   87–101
right-handed frame   119, *119*

sarcolemma   3
sarcomere   4, *4*
sarcomere force–length relationship   28,
    28
  striated muscle   30, *31*
sarcomere length
  force–velocity relations   41–2, *42*
  instability   29–30
  length non-unformities   43–5

sartorius 160
semimembranosum 159
  frog, working range 31
semitendinosus 159
  frog 31
series arrangement 88, *88*
shapes, muscle 24–5
shortening-hold-stretch *47, 48*
singularity 180
size principle of motor unit
  recruitment 11
sliding filament phenomenon 77, 101
small-displacement theory 122
softening and stability 96–101
soleus muscles 8
spatial position 120
stability
  definition 30
  and softening 96–101
static analysis of skeletal muscle, program
  for 146–8, 187–95
  cat medial gastrocnemius
    muscle 148–50, 195–199,
    199–201
stiffness constant 87–8
straight-line models (SLMs) 24
stress 125
stretch–hold shortening *47, 48*
structural models 101–9
structure, muscle 3–8, *4*
submaximal contraction 9
submaximal nerve stimulation 35–6
supramaximal contraction 10
synaptic cleft 13

T-tubules 13, *14*
tension 126
thermal conductors 127
thermal insulators 127
thermodynamic consistency 127
thick filament 5, *5, 7*
thin filament 5–6, *6, 7*

tibiofemoral joint 160
time-dependent modelling 150–1
  analysis of skeletal muscle, program
    for 151–2, 215–25
  cat medial gastrocnemius muscle,
    program for 152, 225–33
  explicit method 202
  implicit method 202, 206–12
titin 7–8
topology 115, *116*
tropomyosin 5
troponin 5, 14
troponin C 6
troponin I 6
troponin T 6
twitch duration 9, *10*
twitch interpolation technique 34–5
two-degrees-of-freedom
  system 212–15, *212*

unfused tetanic contractions 9
unipennate muscles 8, *8*, 25
unstretched length 87

variations 123
virtual displacements 123–4, 128
virtual work
  computer code for 181–4
  identity 177
  obtaining residuals and derivatives
    from 177–8
  principle of 128–31
  see also Newton–Raphson method
viscoelastic behaviour 127
viscous constant 88
Voigt body 88, *109*
voluntary contractions 9

Wilson's θ-method 207, 209–12, 213
working range of the muscle 27

Z-lines 4, 5